Edward James Stone

The Cape Catalogue of 1159 Stars

Edward James Stone

The Cape Catalogue of 1159 Stars

ISBN/EAN: 9783337027100

Printed in Europe, USA, Canada, Australia, Japan

Cover: Foto ©berggeist007 / pixelio.de

More available books at **www.hansebooks.com**

THE CAPE CATALOGUE

OF

1159 STARS,

DEDUCED FROM OBSERVATIONS

AT THE

ROYAL OBSERVATORY, CAPE OF GOOD HOPE,

1856 TO 1861,

REDUCED TO THE EPOCH

1860.

UNDER THE SUPERINTENDENCE OF

E. J. STONE, M.A., F.R.S., F.R.A.S.,
(LATE FELLOW OF QUEEN'S COLLEGE, CAMBRIDGE),

HER MAJESTY'S ASTRONOMER AT THE CAPE.

PUBLISHED BY ORDER OF THE BOARD OF ADMIRALTY, IN OBEDIENCE TO HER MAJESTY'S COMMAND.

CAPE TOWN:
SAUL SOLOMON & CO., 49 & 50, ST. GEORGE'S-STREET.
1873.

CAPE TOWN :

SAUL SOLOMON AND CO., PRINTERS,

49 AND 50, ST. GEORGE'S-STREET.

CAPE CATALOGUE OF 1159 STARS,

FOR THE EPOCH 1860.

The Royal Observatory, Cape of Good Hope, was established by an Order in Council, dated 1820, October 20. The leading idea was to establish a first class Observatory in the Southern Hemisphere, for work of a similar character to that of the Greenwich Observatory in the Northern Hemisphere. The observations were to be made with instruments of the same class, and the results were to be drawn up in the same form, in order that the whole might constitute two corresponding series capable of comparison in all their parts. No opportunity of making observations capable of improving the Theory of Refraction was to be neglected.

The Observatory Buildings were completed ; the instruments, similar to those at that time in use at Greenwich, mounted ; and observations commenced in 1829. But the hand of death was almost upon the first Director. The Assistant, Capt. Ronald, broke down, and left in 1830; and Fallows, after struggling on, somewhat hopelessly, as best he could with the aid of his wife, died in July, 1831, at the early age of 43, with the expectations which had drawn him to South Africa unfulfilled. Much could not have been done under the circumstances in which Fallows was placed, and not much was done. A Catalogue of approximate places of the principal stars, south of the zenith of the Cape, made with small instruments, in Cape Town, was published in the Philosophical Transactions for 1824, and another Catalogue, formed from observations made at this Observatory 1829-1831, appears in the Memoirs of the Royal Astronomical Society, Vol. XIX, published in 1851. It contains the Right Ascensions of 425 Stars : but of these only 88 have corresponding observations in North Polar Distance ; a considerable portion of these places are those of well-known Stars observed for clock error.

Fallows was succeeded by Mr. Thos. Henderson, who remained at the Cape a little more than a year; but that year must ever be a memorable one in the annals of the Cape Observatory. Henderson discovered the sensible parallax of α Centauri, and determined its amount with an accuracy which has left to his successors little more than a verification of his result. He reduced and published, in 1835, the declinations of 172 Stars, and in 1844, in the Memoirs of the Royal Astronomical Society, the Right Ascensions of 174 of the principal Stars, the results of the observations made by him and his assistant at the Cape in 1831 and 1832. It is impossible to over-estimate the value of these papers as affording accurate places of a limited number of Stars at the epoch : but the extent of the Catalogue is small. Henderson removed all the records of his observations from the Observatory, with a view to their subsequent reduction. It is desirable that the original records should be returned for preservation at this Observatory. Mr. Henderson was succeeded by Mr. (now Sir Thomas) Maclear, who arrived at the Cape on the 5th of January, 1834. The observations made by Mr. Maclear and his Assistant in the year 1834 were published and distributed for the use of Astronomers, and some advance was made in the printing of the observations made in 1835, 1836, 1837 ; but the reductions of these years were never completed, and the printing was stopped. It would appear that attention was called off from attempts to form a Star Catalogue to the measurement of an arc of meridian. The field work of this arc was finished in 1847 ; but it was not until 1866 that the volumes containing the results were completely printed and published, and Sir Thos. Maclear placed in a position to receive the congratulations of his contemporaries on the completion of his work. Of the value of this work there cannot be two opinions : but it was allowed to disorganize the other work of the Observatory to such an extent that, when I assumed the Directorship in 1870, I found myself, with a very limited staff, unexpectedly confronted with the results of 36 years of miscellaneous observing in all stages of reduction, nothing completed, and nothing which could be brought forward for publication and use without a very considerable expenditure of time and skilled labour. I fear the course pursued of continuous miscellaneous observing without reduction, has not tended to the advancement of accurate Astronomy to any extent proportional to the labour expended upon the work, and still required to be expended upon it before the results can be rendered useful. Such observing is rarely conducted in a way to facilitate the subsequent reductions or to economize labour in observing. This will be apparent to any one who will count the number of observations of Stars between 67° and 117° North Polar Distance and consider that a Catalogue formed from the results of other years would contain observations of these Stars to very nearly the same relative extent. Of the large number of observations accumulated here from 1834 to 1855, with the Transit instrument and Mural Circles, the places of the Southern Stars, out of the reach of the Northern Observatories, will when reduced, still be of value for proper motions ; but the immense number of observations of well-known Stars made here with the old instruments can now, I fear, never repay the labour required for their reduction.

The Right Ascensions of those Stars which have been used for clock error can do little more than reproduce the assumed tabular places employed in the reductions, and the Right Ascensions of other Stars not further from the Equator than those of the usual clock-star list can never differ much from the results of the Northern Observatories.

The North Polar Distances of the same well-observed Stars can now be of little value. The results are not likely to be compared with those of the Northern Observatories for a discussion of the errors of the refraction tables when results made with more powerful instruments are available.

I have made these remarks, not only in justice to the present staff, and to explain the work upon which they have been employed, but because it was these considerations which led me to pass over the earlier observations, and to commence the systematic reductions with the year 1856, when the Transit Circle was first brought into regular use. I felt that these reductions could not be any longer delayed without the value of the results being greatly diminished. I had, and still have, hopes, that the data collected in the present Catalogue for corresponding observations at the Northern Observatories would be found sufficient to meet the actual requirements of Astronomers, so far as these requirements can be met by the material collected here, and that I might be relieved from the laborious and somewhat useless task of completing the reductions of the earlier observations of Stars whose positions have been fixed already with an accuracy greater than could be expected to be attained in the observations made with the, comparatively speaking, inferior instruments in use at this Observatory before the introduction of the Transit Circle.

The present Catalogue has been formed from the volumes of results of observations made at this Observatory in the years 1856, 1857, 1858, 1859, and 1860, the reductions for which have been completed and the results published since I took charge of the Observatory work in October, 1870. The observations of Stars near the South Pole observed in 1861 have also been included. The whole of the observations combined for the formation of this Catalogue were made with the Transit Circle, an instrument similar in all respects to the Greenwich instrument which has been in use since 1851. The results of the observations made at Greenwich, 1854 to 1860, have been formed into a Catalogue reduced to the epoch 1860. There exist two other Catalogues, of great value, reduced to the same epoch. January 1, 1860, has therefore been chosen as the epoch of the present Catalogue. This choice of epoch affords great facilities for a comparison between the Greenwich and Cape results, and also for the formation of a more general Catalogue, for the whole heavens, by the combination of the different Catalogues reduced to that epoch.

Comparison of the Mean Right Ascensions of Clock-stars in the Greenwich Catalogue for 1860, with those contained in the Cape Catalogue of 1159 Stars for 1860.

Star's Name.	Mean R. A.	No. of G. Obs.	Seconds of Mean R.A. Greenwich.	No. of Cape Obs.	Seconds of Mean R.A. Cape.	Diff. G.—C.
	h m					
α Andromedæ	0' 1	59	9'44	4	9'36	+0'08
γ Pegasi....................	0' 6	67	1'79	15	1'82	—0'03
12 Ceti....................	0'22	42	53'64	6	53'67	—0'03
β Ceti	0'36	49	33'59	44	33'60	—0'01
δ Piscium	0'41	11	25'27	5	25'27	0'00
ε Piscium..................	0'55	74	40'79	7	40'79	0'00
ε Piscium..................	1' 1	17	9'57	2	9'64	—0'07
θ Ceti	1'17	69	1'53	15	1'53	0'00
η Piscium	1'23	74	59'78	6	59'81	—0'03
π Piscium	1'29	12	40'83	3	40'89	—0'06
ν Piscium	1'34	62	8'87	4	8'91	—0'04
o Piscium......	1'38	11	0'28	1	0'36	—0'08
β Arietis	1'46	52	54'72	13	54'71	+0'01
α Arietis	1'59	75	17'28	13	17'26	+0.02
67 Ceti....................	2'10	40	0'10	10	0'11	—0'01
ξ² Ceti	2'20	50	43'13	6	43'19	—0'06
γ² Ceti....................	2'36	56	2'95	10	2'97	—0'02
ε Arietis	2'51	10	12'77	18	12'77	0'00
α Ceti	2'54	57	57'83	13	57'80	+0'03
δ Arietis	3' 3	51	37'74	8	37'74	0'00
17 Tauri	3'36	20	34'10	5	34'11	—0'01
η Tauri	3'39	67	10'09	8	10'07	+0'02
γ¹ Eridani	3 51	40	29'90	13	29'95	—0'05
o¹ Eridani.................	4' 5	10	1'96	14	2'02	—0'06
ε Tauri....................	4 20	67	26'71	11	26'70	+0'01
α Tauri....................	4'27	123	53'43	26	53'41	+0'02
ε Leporis	4'59	23	32'10	11	32 13	—0'03
β Orionis.................	5' 7	101	48'63	42	48'63	0'00
β Tauri	5'17	68	26'67	17	26'64	+0'03
δ Orionis	5'24	38	51'30	15	51'31	—0'01

Star's Name.	Mean R. A.	No. of G. Obs.	Seconds of Mean R.A. Greenwich.	No. of Cape Obs.	Seconds of Mean R.A. Cape.	Diff. G.—C.
	h m		″		″	″
α Leporis....................	5·26	15	33·36	1	33·36	0·00
ε Orionis	5·29	28	6·60	3	6·55	+0·05
ζ Tauri	5·29	12	16·82	4	16·78	+0·04
α Columbæ	5·34	21	34·79	21	34·81	—0·02
α Orionis...................	5·47	86	35·59	28	35·55	+0·04
ν Orionis...................	5·59	12	34·71	8	34·68	+0·03
η Geminorum	6· 6	35	25·65	7	25·58	+0·07
κ Aurigæ	6· 6	18	27·45	3	27·47	—0·02
μ Geminorum	6·14	48	29·43	11	29·42	+0·01
γ Geminorum	6·29	25	37·41	9	37·41	0·00
ε Geminorum	6·35	12	19·00	2	18 91	+0·09
ε Canis Majoris	6·53	28	7·45	28	7·39	+0·06
γ Canis Majoris..........	6·57	15	25·47	6	25·50	—0·03
δ Geminorum	7·11	59	45·56	18	45·53	+0·03
ι Geminorum	7·17	16	1·67	4	1·62	+0·05
α² Geminorum	7·25	61	39·73	2	39·66	+0·07
α Canis Minoris	7·31	118	58·34	30	58·31	+0·03
β Geminorum	7·36	119	44·68	3	44·62	+0·06
6 Cancri	7·54	49	54·86	2	54·80	+0·06
15 Navis	8· 1	25	34·94	11	34·92	+0·02
ψ² Cancri...................	8· 2	14	0·96	2	0·92	+0·04
η Cancri	8·24	40	36·48	7	36·46	+0·02
γ Cancri	8·35	35	10·76	2	10·76	0·00
ε Hydræ	8·39	57	21·57	8	21·56	+0·01
83 Cancri	9·11	56	9·77	7	9·70	+0·07
α Hydræ	9·20	49	42·44	17	42·44	0·00
ε Leonis	9·37	49	53·92	3	53·90	+0·02
ν Leonis	9·50	12	41·29	2	41·29	0·00
π Leonis	9 52	49	48·75	13	48·75	0·00
α Leonis	10· 0	118	54·78	32	54·77	+0·01
γ¹ Leonis	10·12	28	14·95	5	14·97	—0·02
ρ Leonis...................	10·25	58	26·22	12	26·27	—0·05
l Leonis	10·41	60	53·74	6	53·77	—0·03
c Leonis.	10·53	12	29·27	1	29 34	—0·07
χ Leonis ,..................	10·57	44	47·60	16	47·58	+0·02

Star's Name.	Mean R. A.	No. of G. Obs.	Seconds of Mean R A, Greenwich.	No. of Cape Obs,	Seconds of Mean R.A. Cape.	Diff. G. — C.
	h m		s		s	s
δ Leonis	11· 6	54	39·50	4	39·48	+0·02
δ Crateris...................	11·12	33	20·61	15	20·60	+0·01
σ Leonis	11·13	40	54·98	7	54·98	0·00
τ Leonis	11·20	45	44·20	6	44·21	—0·01
υ Leonis	11·29	42	46·84	5	46·87	—0·03
β Leonis	11·41	53	54·98	4	54·92	+0·06
β Virginis	11·43	22	24·13	5	24·12	+0·01
π Virginis.................	11·53	44	41·89	1	41·94	—0·05
ε Corvi	12· 2	19	55·79	13	55·78	+0·01
η Virginis	12·12	55	44·64	5	44·70	—0·06
β Corvi ...., ...	12·27	30	2·34	35	2·35	—0·01
χ Virginis	12·32	21	1·39	2	1·39	0·00
ψ Virginis	12·47	17	4·51	4	4·63	—0·12
δ Virginis.................	12·48	80	33·16	1	33·09	+0·07
θ Virginis.................	13· 2	29	42·23	6	42·21	+0·02
a Virginis	13·17	223	49·27	23	49·27	+0·00
ζ Virginis	13·27	60	33·70	3	33·65	+0·05
m Virginis	13·34	10	16·03	1	16·08	—0·05
η Boötis.....................	13.48	57	1·13	..	..	..
τ Virginis..................	13·54	50	31·42	4	31·48	—0·06
κ Virginis	14· 7	10	41·57	..	..	..
a Boötis	14· 9	130	16·62	14	16·60	+0·02
λ Virginis	14·11	15	32·39	1	32·26	+0·13
α Libræ	14·43	58	8·31	18	8·35	—0·04
20 Libræ	14·55	15	53·01	3	53·04	—0·03
ψ Boötis	14·58	35	26·85	4	26·77	+0·08
β Libræ	15· 9	63	28·62	13	28·64	—0·02
a Coronæ Borealis	15·28	92	45·69	4	45·61	+0·08
a Serpentis	15·37	86	22·44	7	22·42	+0·02
δ Scorpii	15·52	30	3·66	1	3·55	+0·11
β Scorpii	15·57	40	18·08	26	18·10	—0·02
δ Ophiuchi	16· 7	71	0·68	9	0·72	—0·04
σ Scorpii	16·12	11	41·00	6	41·02	—0·02
α Scorpii	16·20	50	49·70	41	49·69	+0·01
κ Ophiuchi	16·51	62	2·56	6	2·53	+0·03

Star's Name.	Mean R. A.	No. of G. Obs.	Seconds of Mean R.A. Greenwich.	No. of Cape Obs.	Seconds of Mean R.A. Cape.	Diff. G.—C.
	h m		s		s	s
α Herculis	17· 8	57	15·92	2	15·91	+0·01
θ Ophiuchi	17·13	43	24·86	23	24·84	+0·02
d Ophiuchi	17·18	14	25·10	16	25·04	+0·06
α Ophiuchi	17·28	105	26·22	11	26·22	0·00
μ Herculis	17·40	54	58·86	6	58·77	+0·09
μ Sagittarii	18· 5	81	23·44	29	23·45	—0·01
δ Sagittarii	18·12	12	1·88	8	1·86	+0·02
λ Sagittarii	18·19	22	19·84	10	19·83	+0·01
α Lyræ........................	18·32	92	11·92	13	11·85	+0·07
φ Sagittarii	18·36	11	54·46	15	54·52	—0·06
β Lyræ........................	18·44	124	54·71	10	54·73	—0·02
σ Sagittarii	18·46	10	34·95	26	34·95	0·00
ζ Aquilæ.....................	18·58	117	58·52	19	58·55	—0·03
ω Aquilæ	19·11	68	14·69	16	14·69	0·00
δ Aquilæ	19·18	61	26·31	21	26·34	—0·03
λ² Sagittarii	19·28	27	11·01	25	11·04	—0·03
γ Aquilæ	19·39	115	36·22	9	36·15	+0·07
α Aquilæ	19·43	159	57·12	42	57·10	+0·02
β Aquilæ.....................	19·48	85	26·14	9	26·17	—0·03
c Sagittarii	19·54	50	2·64	10	2·66	—0·02
a² Capricorni	20·10	32	17·03	26	17·05	—0·02
β Capricorni	20·13	23	8·52	1	8·39	+0·13
ρ Capricorni...............	20·20	20	52·26	20	52·30	—0·04
ψ Capricorni	20·37	17	48·01	9	48·03	—0·02
32 Vulpeculæ	20·48	65	35·66	4	35·67	—0·01
θ Capricorni ,.............	20·58	41	4·34	6	4·35	—0·01
ζ Cygni	21· 6	84	58·75	3	58·67	+0·08
ι Capricorni	21·14	48	26·78	14	26·76	+0·02
β Aquarii....................	21·24	57	11·15	38	11·15	0·00
γ Capricorni...............	21·32	16	19·77	15	19·78	—0·01
ε Pegasi	21·37	66	18·60	10	18·60	0·00
δ Capricorni...............	21·39	26	18·53	12	18·54	—0·01
16 Pegasi...................	21·46	50	41·63	4	41·65	—0·02
α Aquarii...................	21·58	70	35·50	12	35·49	+0·01
ι Aquarii	21·58	11	52·29	6	52·37	—0·08

Star's Name.	Mean R. A.	No. of G. Obs.	Seconds of Mean R.A. Greenwich.	No. of Cape Obs.	Seconds of Mean R.A. Cape.	Diff. G.—C.
	h m		"		"	"
θ Aquarii	22˙ 9	49	26˙59	15	26˙61	—0˙02
σ Aquarii	22˙23	14	14˙11	6	14˙06	+0˙05
η Aquarii	22˙28	54	9˙65	11	9˙70	—0˙05
ζ Pegasi	22˙34	49	28˙81	3	28˙84	—0˙03
λ Aquarii	22˙45	35	18˙48	3	18˙50	—0˙02
α Piscis Australis	22˙49	59	54˙38	40	54˙36	+0˙02
α Pegasi.....................	22˙51	54	47˙33	3	47˙30	+0˙03
γ Piscium	23˙ 9	60	54˙47	11	54˙47	0˙00
κ Piscium..................	23˙19	56	45˙31	7	45˙34	—0˙03
ι Piscium	23˙32	69	44˙99	4	45˙09	—0˙10
δ Sculptoris...............	23˙41	17	37˙60	11	37˙62	—0˙02
ω Piscium	23˙52	79	7˙39	12	7˙40	—0˙01

This list contains the Stars whose Right Ascensions have been employed for the determination of clock error in the reduction of the observations included in the present Catalogue. The Mean Right Ascensions adopted for the determination of clock error have been those of the Greenwich Catalogue of 2022 Stars for 1860, and the results have been carried back to the year of observation with the precessions, secular variations, and proper motions given in that Catalogue. No correction for epoch has been applied, and the fundamental epoch of this Catalogue should therefore be that of the Greenwich Catalogue. The differences between the Right Ascensions adopted for clock error and the resulting Right Ascensions are generally small, and the mean difference is only +0^s·002. But the differences are not quite constant throughout the twenty-four hours. The following are the Mean Differences for intervals of three hours :

GREENWICH R.A.—CAPE R.A.

h h 0 to 3	h h 3 to 6	h h 6 to 9	h h 9 to 12	h h 12 to 15	h h 15 to 18	h h 18 to 21	h h 21 to 24
—0˙007	+0˙005	+0˙030	+0˙003	—0˙007	+0˙010	—0˙006	—0˙005

The only one of these differences which is very strongly marked is that from 6^h to 9^h, but the Greenwich Right Ascensions from 5^h to 9^h, both inclusive, are systematically greater than the corresponding Cape

Right Ascensions, the mean difference being +0ˢ·021. These Stars were generally observed at the Cape during the months of December, January, and February. The changes of temperature in the evening are exceedingly rapid, the average change of temperature in January between 6ʰ and 12ʰ being 5°·4 F. I think it probable that, so far as the above differences indicate errors in the Cape places, they may be attributed to the mean rate not being strictly applicable over a group of Stars extending from soon after sunset to late on in the night. The Transit Clock was not, and is not now, protected from rapid changes of temperature, and the compensation most probably lags behind in its action.

The North Polar Distances have been reduced with the refractions of the Tabulæ Regiomontanæ to 85° Z.D. ; but below 85° Z.D., instead of introducing the break in the mean refractions recommended by Bessel, the tables have been extended by using the mean refractions from the Fundamenta multiplied by the constant required to bring up the mean refractions of the Fundamenta to those of the Tabulæ Regiomontanæ.

The latitude of the Observatory assumed in the volumes of results has been that determined by Henderson, 33° 56′ 3″·2 South.

In the formation of this Catalogue I have adopted, from a preliminary investigation, 33° 56′ 3″·56 South latitude.

The following table exhibits the corrections which this adopted quantity would still appear to require with the refraction tables in use. The weights in this table have been calculated from the probable errors at different zenith distances, given in my paper in the monthly notices of the Royal Astronomical Society :

LATITUDE INVESTIGATION.

Star's Name.	No. of Obs.		Weight.	Excess of N.P.D. (Below — above.)	
	Above Pole.	Below Pole.			
o Octantis	19	27	22·23	+0·11	=2x−169y
β Hydri.....................	110	92	93·88	+0·11	=2x−194y
Lacaille 634	3	5	3·77	+0·15	=2x−173y
γ Hydri	27	25	24·27	−0·68	=2x−216y
δ Mensæ	7	8	7·41	+0·09	=2x−184y
Lacaille 1639	9	4	5·00	+0·06	=2x−180y
Lacaille 1752	10	7	7·43	−0·31	=2x−213y
A Octantis	14	10	11·33	+0·38	=2x−172y
β Argûs.....................	9	4	3·43	+1·15	=2x−280y
ζ Octantis	17	3	3·93	+1·94	=2x−173y

Star's Name.	No. of Obs. Above Pole.	No. of Obs. Below Pole.	Weight.	Excess of N.P.D. (Below — above.)
✳ Octantis	34	25	28·45	+0·23 =2x—169y
B.A.C. 4460	6	2	2·77	—1·14 =2x—172y
κ Octantis	17	11	12·80	—0·76 =2x—172y
ε Apodis	4	1	1·31	+1·98 =2x—188y
Z Octantis	53	25	32·38	—0·08 =2x—171y
B.A.C. 4883	5	3	3·45	+1·28 =2x—179y
ρ Octantis	19	18	18·12	+0·03 =2x—176y
Brisbane 5607...............	16	6	8·21	+0·17 =2x—173y
γ Apodis	4	4	3·81	+0·65 =2x—193y
β Apodis	3	4	3·47	+0·60 =2x—197y
ι Apodis	3	2	1·74	—1·92 =2x—266y
Brisbane 6058	16	9	11·24	—0·19 =2x—170y
σ Octantis	45	41	42·49	—0·29 =2x—169y
ε Pavonis.....................	2	5	3·29	+0·78 =2x—227y
σ Pavonis.....................	4	5	3·63	—1·89 =2x—278y
B Octantis....................	33	31	31·71	+0·41 =2x—169y
C Octantis	22	35	27·27	—0·43 =2x—171y
β Octantis	1	5	1·92	+0·42 =2x—180y
τ Octantis	39	87	54·04	—0·05 =2x—176y
γ² Octantis	1	5	1·88	—1·21 =2x—177y

In these equations it is assumed that the true refraction = Tabular $(1—y)$.

South latitude = 33° 56′ 3″.56 + z.

From them we find, for the South latitude of the Observatory,—

33° 56′ 3″·55 + 91″·y.

There are not a sufficient number of Stars, far removed from the pole, observed at their upper and lower culminations in North Polar Distance to allow of any accurate determination of y from the results of this Catalogue.

The latitude of the Cape Observatory must therefore be still considered as uncertain to half a second. If the value of y be determined from these equations, the results are,—

$$z = + 0″·416. \quad y = + 0·0047.$$

The following method has been employed in the formation of the Catalogue :

In the first place, it must be remarked that the observations contained in the separate volumes of results are not corrected for the proper motion for the fraction of the year of observation except for the Nautical Almanac

Stars, for which Stars it has been included in the Star correction. All the observations of Stars common to this Catalogue and the Greenwich Catalogue for 1860, which are not contained in the Nautical Almanac, and for which proper motions are given in the Greenwich Catalogue, were first corrected for the proper motion for the fraction of a year, and then all the Stars common to the Greenwich Catalogue brought up to the epoch 1860, January 1, from the different years with the precessions, secular variations, and proper motions of the Greenwich Catalogue. The places for the Stars not contained in the Greenwich Catalogue were then brought up to 1860, with the precessions for the year of observation, or mean precessions, or the precessions approximately corrected for secular variations from the British Association Catalogue, and very approximate places for 1860 thus found. With these places the precessions and secular variations, for 1860, were computed. The results for the several years were then brought up to 1860 with greater accuracy, and combined in proportion to the number of observations in each year for places which were so far uncorrected for proper motions, except for the Nautical Almanac Stars to which they were applied.

A comparison was then made between these results and the places of the same Stars given in the following Catalogues:

Brisbane, in N. P. D. only, Fallows, Johnson, Henderson, and an unfinished Cape Catalogue for 1840, formed from the observations made, 1834 to 1840, the reductions for which were examined and completed so far as necessary for this comparison. From these comparisons proper motions were adopted for use in the present Catalogue. These proper motions were then applied, for the Stars not in the Greenwich Catalogue or Nautical Almanac, for the interval between the mean time of observation and the epoch of the Catalogue 1860, January 1. In cases where the ·adopted proper motions for Nautical Almanac Stars differed from those used in the Nautical Almanac the necessary corrections were applied. The whole of the results are therefore reduced with the proper motions given in the Catalogue, and the corrections for proper motion for the fraction of the year of observation have been applied in all cases when required. The intervals between this Catalogue and those compared with it for proper motions are but small, and no extreme accuracy can therefore be expected in the resulting proper motions; but after the experience of Baily, and some further trials of my own, it has not appeared to me desirable to trust to Lacaille's places for proper motions, and these adopted proper motions are therefore probably the best available. I believe them sufficiently accurate to bring up the places of the present Catalogue for twenty years with at least the accuracy of the earlier Catalogues.

I have not thought it necessary to give the Star constants. Very approximate constants computed for the reductions will be found for the Southern Stars in the volumes of results for the several years, 1856—1860, and the constants for the Greenwich Stars are given for 1860 in the Greenwich Catalogue. The secular variations have been corrected for the change of m and n, as well as for changes in place. In some cases, where southern stars have been observed in North Polar Distance but no corresponding observations in Right Ascensions have been available during the period 1856 to 1861, I have reduced observations made in other years to fill up the gaps. These cases are clearly marked by the mean year of observation and by a note.

The references to other Catalogues have been much restricted for want of room. The Greenwich Catalogue generally referred to is that for 1860. When the places of a star are not contained in the Greenwich 1860 Catalogue, but are contained in the Greenwich Catalogue for 1864, references are given to the latter Catalogue and distinguished by an *. Henderson's Catalogue is only referred to for stars not contained in the Greenwich Catalogues. The letter H is prefixed to Henderson's numbers. Fallows' and Johnson's Catalogues are distinguished by the letter J. being prefixed to Johnson's Catalogue. In a few cases references have been made to the Radcliffe Catalogue for 1860, under the letter R. During the time this Catalogue has been in preparation the Observatory Staff, which should consist of four assistants, has consisted of but two. The first assistant, Mr. Mann, was absent from serious illness, which led to his resignation, and, very shortly afterwards, death. The third assistantship has been vacant for three years owing to deaths and delays in filling up the appointment. Under such circumstances, a very considerable portion of the mere arithmetical work has fallen directly upon me, and all of the examination. I have, however, when the computations have been made by me, re-computed the work again after an interval, and have taken all the precautions which occurred to me to insure accuracy. I hope that the errors contained in this Catalogue will be found but few, and those not important, and that the Catalogue may be found not unworthy of the Observatory. The reductions of the observations which this Catalogue contains, and the formation of the Catalogue, have occupied my chief thoughts during the three years I have been at the Cape.

It may be mentioned that the present is the first Star Catalogue ever printed at the Cape, and the first of any extent yet published from materials collected at this Observatory since its foundation. The printing has been done by Messrs. Saul Solomon and Co., who, at their own instigation, ordered out a fresh fount of type for the printing of the Cape observations. My thanks are due to them and to their leading printer, Mr. S. Wiid, for the care bestowed in passing through the press a very heavy work, and one of an unusual character in their office.

E. J. STONE.

1873, October 5.

THE
CAPE CATALOGUE OF 1159 STARS

FOR

1 8 6 o,

DEDUCED FROM

OBSERVATIONS

MADE AT THE

ROYAL ÓBSERVATORY, CAPE OF GOOD HOPE,

1856—1861.

No.	No. in B.A.C.	Magnitude	Star's Name.	Mean R.A. 1860, Jan. 1.			Mean year and Fraction of year.	No. of Obs. of R.A.	Annual Precess. in R.A. for 1860.	Secular Variation of Precess. in R.A.	Annual Proper Motion in R.A.
				h	m	s	1800				
1	4	2	21 Andromedæ... α	0	1	9·36	59·08	4	+3·076	+0·0181	+0·009
2	11	4	Phœnicis.......... ε	0	2	17·66	59·86	6	3·058	−0·0292	+0·008
3	19	5	Octantis......... γ³	0	3	36·48	56·44	7	2·901	−0·2140	−0·014
4	26	3·2	88 Pegasi......... γ	0	6	1·82	59·29	15	3·081	+0·0099	0·000
5	35	6	B.F. 3310............	0	7	45	..	..	3·064	−0·0032	..
6	36	6	35 Piscium.........	0	7	46·29	60·89	2	3·078	+0·0066	+0·004
7	64	5	Tucanæ........... ζ	0	12	44·83	60·00	3	2·908	−0·0563	+0·265
8	66	6·5	41 Piscium........ d	0	13	23·73	59·18	9	+3·082	+0·0066	−0·002
9	71	6·7	Octantis... .... o	0	13	24·98	58·13	42	−2·141	+4·7690	+0·040
10	70	4·5	Tucanæ........... π	0	14	7·74	59·74	1	+2·841	−0·0684	..
11	88	3	Hydri............. β	0	18	19·91	59·63	238	2·568	−0·0913	+0·720
12	89	6	45 Piscium.........	0	18	28·95	58·93	10	3·085	+0·0066	−0·001
13	94	2	Phœnicis......... α	0	19	21·44	59·67	1	2·967	−0·0229	+0·022
14	112	6	12 Ceti..............	0	22	53·67	57·08	6	3·061	+0·0008	−0·002
15	127	4	Tucanæ........... β²	0	25	6·52	59·75	3	2·776	−0·0452	+0·008
16	134	5	Lacaille 123........	0	26	19·99	59·85	1	2·761	−0·0446	+0·012
17	143	5	Lacaille 137........	0	27	47·40	59·82	2	2·856	−0·0305	+0·035
18	145	6·5	13 Ceti.............	0	28	2·55	56·55	1	3·060	+0·0012	+0·017
19	149	6	W.B. 0h 484........	0	28	39·86	59·66	2	3·109	+0·0102	..
20	176	5	Lacaille 172........	0	33	50·96	59·74	2	2·728	−0·0362	+0·125
21	183	5	Phœnicis.......... μ	0	34	41·98	59·93	1	2·857	−0·0232	−0·009
22	196	2	16 Ceti............ β	0	36	33·60	59·12	44	3·000	−0·0056	+0·013
23	222	4·5	63 Piscium........ δ	0	41	25·27	58·47	5	3·101	+0·0078	+0·003
24	242	5·6	20 Ceti..............	0	45	51	..	..	3·063	+0·0035	−0·004
25	265	6	Lacaille 259........	0	49	40·38	59·83	1	2·677	−0·0250	+0·001
26	272	5·4	Sculptoris......... α	0	51	51·38	59·95	8	2·898	−0·0102	+0·001
27	288	4	71 Piscium .... ι	0	55	40·79	58·27	7	3·112	+0·0086	−0·002
28	313	6	27 Ceti..............	0	58	36	..	..	3·008	−0·0001	+0·000
29	317	3·4	Phœnicis.......... β	0	59	49·77	60·01	5	2·696	−0·0184	−0·006
30	328	6·5	80 Piscium......... e	1	1	9·64	56·59	2	3·102	+0·0077	−0·021
31	333	5	Tucanæ .... ι	1	1	45·15	60·18	2	2·388	−0·0252	+0·003
32	340	5	Phœnicis........... ζ	1	2	28·98	59·86	2	2·538	−0·0223	−0·021
33	341	6	Piazzi 0h 311.......	1	2	46·30	60·90	2	3·168	+0·0134	+0·018
34	368	5·4	86 Piscium....... ζ¹	1	6	25·16	59·61	7	3·118	+0·0090	+0·008
35	380	4·5	Phœnicis........... ν	1	8	51·63	59·96	7	+2·658	−0·0160	+0·070

No	Mean N.P.D. 1860, Jan. 1.	Mean Year and Fraction of Year.	No. of Obs. of N.P.D.	Annual Precess. in N.P.D. for 1860.	Secular Variation of Precess. in N.P.D.	Annual Proper Motion in N.P.D.	No. for reference.			
							Lacaille.	Brisbane.	Fallows or Johnson.	Greenwich or Henderson.
	° ′ ″	1800		″	″	″				
1	61 40 57·74	57·98	14	− 20·06	+ 0·011	+ 0·15	..	7384	..	3
2	136 31 11·78	59·71	9	20·05	+ 0·013	+ 0·19	9742	3	2.J 1	..
3	173 0 8·85	56·44	7	20·05	+ 0·015	− 0·03	9756	5	J 2	H 8
4	75 35 42·05	59·18	33	20·05	+ 0·021	+ 0·02	..	11	..	10
5	100 20 53·20	59·83	1	20·04	+ 0·023	..	..	..	..	..
6	81 57 24·46	60·89	2	20·04	+ 0·024	+ 0·05	..	..	..	12
7	155 41 52·03	60·00	3	20·03	+ 0·033	− 1·18	40	26	J 4	..
8	82 35 15·10	59·18	9	20·02	+ 0·035	− 0·01	..	..	..	19
9	179 8 29·15	58·14	46	20·02	− 0·009	0 00	..	32	J 5	H 2
10	160 24 9·52	59·74	1	20·02	+ 0·035	+ 0·04	53	29	..	..
11	168 2 34·67	59·37	202	19·99	+ 0·039	− 0·32	74	40	5.J 6	H 14
12	83 4 59·05	58·86	8	19·99	+ 0·045	+ 0·07	..	..	..	24
13	133 3 59·25	59·67	1	19·98	+ 0·046	+ 0·40	87	44	6.J 8	H 59
14	94 43 52·60	58·19	13	19·96	+ 0·053	+ 0·01	..	..	7	31
15	153 43 47·09	59·75	3	19·94	+ 0·053	+ 0·03	119	58	9.J 10	..
16	153 48 9·93	59·85	1	19·92	+ 0·055	− 0·03	123	61	J 12	..
17	143 8 48·78	59·82	2	19·91	+ 0·059	..	137	67	..	..
18	94 21 51·19	56·55	1	19·91	+ 0·063	+ 0·04	..	..	..	38
19	77 33 28·59	59·66	2	19·90	+ 0·065	..	..	..	..	40
20	150 14 29·07	59·74	2	19·84	+ 0·067	− 0·54	172	81	..	..
21	136 51 15·22	59·93	1	19·83	+ 0·071	+ 0·12	177	84	J 13	..
22	108 45 20·51	58·81	122	19·80	+ 0·078	− 0·02	..	..	13.J 14	50
23	83 10 39·45	59 02	12	19·73	+ 0·090	+ 0·05	..	..	14	59
24	91 54 19·19	59·86	1	19·66	+ 0·097	+ 0·01	..	..	15.J 17	62
25	143 56 59·68	59·83	1	19·59	+ 0·092	+ 0·09	259	121	..	..
26	120 6 53·81	59·87	7	19·54	+ 0·103	+ 0·03	266	125	16.J 18	69
27	82 51 52·14	59·18	34	19·47	+ 0·118	0·00	..	..	..	72
28	100 43 45·27	59·86	1	19·40	+ 0·120	+ 0·01	..	..	..	133*
29	137 28 8·78	60·01	5	19·38	+ 0·110	+ 0·04	308	145	17.J 19	..
30	85 5 30·88	56·68	3	19·35	+ 0·128	+ 0·19	..	..	..	76
31	152 31 26·22	60·18	2	19·33	+ 0·101	− 0·01	326	155	..	..
32	145 59 42·19	59·84	5	19·32	+ 0·108	+ 0 02	318	156	18.J 21	..
33	75 4 19·46	60·90	2	19·31	+ 0·133	+ 0·17	..	..	..	80
34	83 9 57·64	59·61	7	19·22	+ 0·138	+ 0·07	..	..	..	82
35	136 16 49·05	59·92	9	− 19·16	+ 0·123	− 0·15	337	172	..	..

No.	No. in B.A.C.	Magnitude	Star's Name.	Mean R.A. 1860, Jan. 1.	Mean Year and Fraction of Year.	No. of Obs. of R.A.	Annual Precess. in R.A. for 1860.	Secular Variation of Precess. in R.A.	Annual Proper Motion in R.A.
				h m s	1800		s	s	s
36	..	7	Lalande 2312........	1 10 8·05	60·92	3	+ 3·169	+ 0·0130	..
37	392	5	Tucanæ............ κ	1 11 0·74	59·77	1	1·975	− 0·0157	+ 0·081
38	398	5	Lacaille 361........	1 12 11·56	60·22	2	2·090	− 0·0181	+ 0·001
39	420	3	45 Ceti.......... θ	1 17 1·53	59·39	15	3·003	+ 0·0018	− 0·007
40	422	5	Lacaille 391........	1 17 7·28	60·22	2	2·026	− 0·0149	− 0·017
41	426	5	Lacaille 392........	1 18 28·70	59·84	3	2·665	− 0·0125	− 0·001
42	427	5	93 Piscium...... ρ	1 18 42·77	59·78	2	3·222	+ 0·0163	− 0·003
43	431	5	94 Piscium	1 19 8·35	59·93	2	3·224	+ 0·0163	+ 0·004
44	447	3	Phœnicis.......... γ	1 22 16·89	59·90	4	2·618	− 0·0127	− 0·004
45	448	5	98 Piscium...... μ	1 22 51·15	56·71	2	3·117	+ 0·0094	+ 0·019
46	453	4·3	99 Piscium ...... η	1 23 59·81	59·21	6	3·197	+ 0·0141	+ 0·000
47	461	4	Phœnicis.......... δ	1 25 25·00	59·89	2	2·496	− 0·0141	+ 0·009
48	476	6	101 Piscium........	1 28 17·56	59·79	10	3·197	+ 0·0138	+ 0·002
49	477	6	Piazzi 1ʰ 120......	1 28 20·56	60·90	2	3·223	+ 0·0155	+ 0·011
50	488	6	102 Piscium...... π	1 29 40·89	56·98	3	3·175	+ 0·0124	− 0·007
51	507	1	Eridani............ α	1 32 29·68	59·61	15	2·233	− 0·0130	+ 0·008
52	518	5·4	106 Piscium...... ν	1 34 8·91	59·59	4	3·117	+ 0·0090	− 0·004
53	537	4	110 Piscium ...... ο	1 38 0·36	56·78	1	3·154	+ 0·0110	+ 0·006
54	539	6	Piazzi 1ʰ 167........	1 38 58	..	..	3·009	+ 0·0039	..
55	541	5	Sculptoris .... ε	1 39 5·20	59·92	3	2·802	− 0·0039	+ 0·009
56	550	5	Eridani............ g²	1 40 45·72	59·86	3	+ 2·282	− 0·0109	+ 0·012
57	554	5·6	Hydri................	1 41 19·86	56·45	4	− 0·125	+ 0·1722	− 0·012
58	557	6	Octantis..........	1 41 51·67	56·42	1	− 2·124	+ 0·5901	+ 0·091
59	565	3	55 Ceti............ ζ	1 44 43	..	..	+ 2·957	+ 0·0021	− 0·002
60	572 573	4·3	5 Arietis.......... γ	1 45 51·13	58·73	1	+ 3·273	+ 0·0172	+ 0·002
61	584	6	Lacaille 634........	1 45 55·72	56·54	8	− 4·462	+ 1·3294	+ 0·026
62	577	3·2	6 Arietis.......... β	1 46 54·71	58·92	13	+ 3·292	+ 0·0182	+ 0·002
63	582	5	Lacaille 559........	1 48 1·74	59·89	3	2·421	− 0·0091	− 0·015
64	585	5	Phœnicis.......... φ	1 48 33·13	59·86	1	2·500	− 0·0083	− 0·015
65	592	6	8 Arietis.......... ι	1 49 42·58	57·73	3	3·262	+ 0·0163	+ 0·007
66	594	6	56 Ceti............	1 50 6	..	..	2·807	− 0·0021	..
67	596	4	Eridani............ χ	1 50 30·25	59·82	2	2·270	− 0·0088	+ 0·067
68	607	6	Piazzi 1ʰ 222.......	1 51 49·80	59·78	4	3·305	+ 0·0185	+ 0·017
69	623	3	Hydri.., ... α	1 54 21·39	59·91	21	1·856	− 0·0025	+ 0·034
70	633	6	60 Ceti............	1 56 1	..	..	+ 3·066	+ 0·0072	+ 0·008

No.	Mean N.P.D. 1860, Jan. 1.	Mean Year and Fraction of Year.	No. of Obs. of N.P.D.	Annual Precess. in N.P.D. for 1860.	Secular Variation of Precess in N.P.D.	Annual Proper Motion in N.P.D.	No. for reference.			
							Lacaille.	Brisbane.	Fallows or Johnson.	Greenwich or Henderson.
	° ′ ″	1800		″	″	″				
36	76 29 44·39	60·92	4	− 19·12	+ 0·147	··	··	··	··	··
37	159 37 12·71	59 77	1	19·10	+ 0·096	− 0·12	356	178	··	··
38	157 8 13·10	60·22	2	19·07	+ 0·102	− 0·01	361	180	··	··
39	98 54 24·78	59·14	42	18 93	+ 0·153	+ 0·22	··	··	20.J 22	92
40	157 7 1·61	60·22	2	18·93	+ 0·106	+ 0·03	391	196	··	··
41	132 13 20·09	59·84	3	18·89	+ 0·139	+ 0·06	392	199	J 23	··
42	71 33 28·78	59·78	2	18·89	+ 0·166	− 0·06	··	··	··	94
43	71 29 10·79	59·93	2	18·87	+ 0·167	+ 0·01	··	··	··	95
44	134 2 11·42	59·90	4	18 78	+ 0·142	+ 0·24	419	209	22.J 25	H 58
45	84 34 44·64	56 71	2	18·76	+ 0·167	+ 0·18	··	··	23	96
46	75 22 37·61	59·43	24	18·72	+ 0·175	0·00	··	··	··	101
47	139 48 4·44	59·89	2	18·68	+ 0·140	− 0·14	440	216	24.J 26	··
48	76 3 20·96	59·50	4	18·59	+ 0·183	− 0·02	··	··	··	104
49	73 17 3·27	60·90	2	18·59	+ 0·185	− 0·07	··	··	··	105
50	78 34 33·96	57·18	4	18·54	+ 0·185	− 0·03	··	··	··	107
51	147 56 56·79	58·43	27	18 44	+ 0·136	+ 0·07	484	239	25.J 27	H 36
52	85 13 20·41	59·53	16	18·39	+ 0·190	+ 0·04	··	··	27	112
53	81 32 54·49	57·73	3	18·25	+ 0·199	− 0·01	··	··	.29	117
54	96 26 6·76	59·94	1	18·22	+ 0·191	··	··	··	··	··
55	115 45 12·88	59·92	3	18·21	+ 0·178	+ 0·08	511	··	30.J 29	··
56	144 13 33·62	59·86	3	18·15	+ 0·150	− 0·04	523	254	··	··
57	169 51 13·85	56·45	4	18·13	− 0·001	+ 0·02	551	259	··	··
58	173 41 12·65	56·42	1	18 11	− 0·125	− 0·06	576	262	··	··
59	101 1 40·67	56·49	1	18·01	+ 0·197	+ 0·12	··	··	31.J 31	H 109
60	71 23 35·44	58·73	1	17·95	+ 0·221	+ 0·11	··	··	··	123 124
61	175 28 30·09	56·53	8	17·95	− 0·283	− 0·01	634	··	··	··
62	69 52 40·64	59·12	30	17·91	+ 0·224	+ 0·11	··	··	··	125
63	136 59 22·89	59·88	5	17·87	+ 0·168	+ 0·15	559	272	··	··
64	133 11 6·31	59·86	1	17·85	+ 0·174	+ 0·04	565	274	J 32	··
65	72 52 2·59	57·73	3	17·80	+ 0·227	− 0·01	··	275	··	126
66	113 12 43·55	59·82	1	17·79	+ 0·196	··	568	··	··	··
67	142 18 24·76	59·82	2	17·77	+ 0·161	− 0·25	575	278	32.J 33	··
68	69 37 23·92	59·78	4	17·72	+ 0·234	+ 0·11	··	··	··	128
69	152 15 7·70	59·89	10	17·61	+ 0·137	− 0·01	605	287	33.J 36	H 25
70	90 32 53·18	59·90	1	− 17·53	+ 0·225	+ 0·02	··	··	··	269*

No.	No. in B.A.C.	Magnitude.	Star's Name.	Mean R.A. 1860, Jan. 1.			Mean Year and Fraction of Year.	No. of Obs. of R.A.	Annual Precess. in R.A. for 1860.	Secular Variation of Precess. in R.A.	Annual Proper Motion in R.A.
				h	m	s					
71	632	6	Piazzi 1^h 243........	1	56	2·19	1800 57·68	2	+ 3·277	+ 0·0167	+ 0·007
72	634	5	Phœnicis.......... χ	1	56	5·48	59·88	4	2·415	— 0·0075	— 0·004
73	648	2	13 Arietis........ a	1	59	17·26	59·12	13	3·352	+ 0·0203	+ 0·012
74	659	6	Lacaille 640........	2	1	37·49	59·83	1	2·078	— 0·0053	..
75	663	6·7	Piazzi 1^h 266......	2	2	24	..	..	3·114	+ 0·0091	..
76	682	5·6	17 Arietis........ η	2	4	58·20	59·67	4	3·332	+ 0·0188	+ 0·009
77	684	4·5	65 Ceti........... ξ¹	2	5	34·97	56·78	1	3·172	+ 0·0115	— 0·004
78	688	5	Fornacis.......... μ	2	6	44·44	59·86	4	2·644	— 0·0032	— 0·003
79	704	6	67 Ceti..............	2	10	0·11	59·50	10	2·983	+ 0·0049	+ 0·003
80	707	6·5	22 Arietis........ θ	2	10	20·69	60·88	3	3·324	+ 0·0179	— 0·002
81	717	4	Eridani.......... φ	2	11	30·31	59·87	5	2·138	— 0·0044	+ 0·005
82	745	5·6	24 Arietis........ ξ	2	17	19·12	56·71	1	3·205	+ 0·0126	+ 0·000
83	747	6	71 Ceti.............	2	17	54	..	..	3·027	+ 0·0066	+ 0·005
84	756	4	Hydri.......... δ	2	19	16·08	59·89	7	1·052	+ 0·0294	— 0·010
85	760	4	73 Ceti........... ξ²	2	20	43·19	59·10	6	3·178	+ 0·0115	+ 0·001
86	763	4·5	Eridani........... κ	2	21	51·10	59·89	6	2·200	— 0·0035	+ 0·000
87	771	6	27 Arietis..........	2	23	8·84	58·31	3	3·312	+ 0·0166	0·000
88	781	5	76 Ceti........... σ	2	25	28	..	..	2·847	+ 0·0024	— 0·001
89	787	6	Lacaille 785.........	2	27	3·75	59·96	5	2·229	— 0·0029	..
90	808	6·5	32 Arietis........ ν	2	30	52·37	59·93	2	3·392	+ 0·0192	— 0·002
91	820	5·6	Horologii.......... η	2	32	47·29	59·95	2	1·969	— 0·0001	..
92	828	5	Lacaille 827.........	2	34	27·69	59·90	3	2·280	— 0·0021	+ 0·006
93	825	6·5	34 Arietis........ μ	2	34	28·75	60·90	1	3·366	+ 0·0179	— 0·001
94	832	4	Eridani........... ι	2	35	8·49	59·91	5	2·358	— 0·0021	+ 0·003
95	837	3·4	86 Ceti........... γ¹	2	36	2·97	59·42	10	3·111	+ 0·0093	— 0·011
96	845	4	87 Ceti.......... μ	2	37	22·71	56·71	1	3·214	+ 0·0124	+ 0·017
97	849	5	Hydri.............. ε	2	37	26·95	59·91	3	0·878	+ 0·0348	+ 0·017
98	864	6	Lacaille 875..........	2	40	16·96	59·98	2	2·257	— 0·0016	— 0·002
99	867	6	40 Arietis...........	2	40	41·52	59·26	4	3·347	+ 0·0168	+ 0·005
100	872	4	41 Arietis...........	2	41	44·86	60·60	2	3·508	+ 0·0228	+ 0·003
101	879	5	Fornacis .... β	2	43	13·87	59 93	9	2·505	— 0·0008	+ 0·008
102	882	5	Hydri. .... ζ	2	43	24·00	59·91	4	0 885	+ 0·0333	+ 0·015
103	911	6	Lacaille 937..........	2	49	18·53	59·96	3	1·268	+ 0·0169	+ 0·001
104	913	6	47 Arietis............	2	50	4·85	59·71	2	3·403	+ 0·0180	+ 0·016
105	921	4·5	48 Arietis.......... ε	2	51	12·77	58·91	18	+ 3·417	+ 0·0184	— 0·001

No.	Mean N.P.D. 1860, Jan. 1.	Mean Year and Fraction of Year.	No. of Obs. of N.P.D.	Annual Precess. in N.P.D. for 1860.	Secular Variation of Precess.in N.P.D.	Annual Proper Motion in N.P.D.	No. for reference.			
							Lacaille.	Brisbane.	Fallows or Johnson.	Greenwich or Henderson.
	° ′ ″	1800		′	″	″				
71	72 25 17·30	57·68	2	− 17·54	+ 0·239	− 0·04	..	290	..	136
72	135 23 19·35	59·88	4	17·54	+ 0·179	+ 0·02	610	291	J 37	..
73	67 12 5·05	59·94	38	17·40	+ 0·251	+ 0·15	..	295	..	137
74	145 45 5·98	59·83	1	17·30	+ 0·161	+ 0·08	640	301	..	..
75	86 25 56·89	59·82	1	17·26	+ 0·237	..	..	..	..	..
76	69 26 54·69	59·63	3	17·15	+ 0·259	− 0·01	..	..	..	141
77	81 48 43·49	56·78	1	17·12	+ 0·248	+ 0·04	..	..	34	142
78	121 22 55·22	59·86	4	17·07	+ 0·210	− 0·08	666	315	..	..
79	97 4 8·98	59·63	28	16·92	+ 0·241	+ 0·14	..	..	..	147
80	70 44 54·73	60·88	3	16·90	+ 0·268	+ 0·01	..	..	..	148
81	142 9 41·37	59·88	7	16·85	+ 0·176	+ 0·05	693	327	35.J 38	..
82	80 1 31·12	56·71	1	16·56	+ 0·271	+ 0·05	..	..	38	153
83	93 24 56·80	59·91	1	16·53	+ 0·257	− 0·06	..	..	..	324*
84	159 17 50·59	59·89	7	16·47	+ 0·095	− 0·01	747	351	39.J 41	..
85	82 10 9·97	59·23	13	16·39	+ 0·274	+ 0·02	..	..	..	157
86	138 20 1·24	59·89	7	16·34	+ 0·193	+ 0·04	753	353	40.J 42	..
87	72 55 2·13	58·31	3	16·27	+ 0·290	+ 0·09	..	..	..	159
88	105 51 39·83	59·90	1	16·15	+ 0·253	+ 0·07	..	..	42.J 43	376*
89	136 29 22·65	59·96	5	16·07	+ 0·202	− 0·11	785	367	..	..
90	68 38 47·22	59·93	2	15·87	+ 0·310	+ 0·02	..	..	..	168
91	143 9 1·50	59·96	3	15·76	+ 0·184	+ 0·02	821	378	..	..
92	133 29 39·85	59·88	4	15·67	+ 0·214	+ 0·03	827	383	J 46	..
93	70 35 13·95	60·90	1	15·67	+ 0·313	+ 0·05	..	..	..	173
94	130 27 22·39	59·91	5	15·64	+ 0·222	+ 0·06	831	384	J 47	..
95	87 21 23·06	59·26	28	15·58	+ 0·292	+ 0·19	..	..	..	178
96	80 28 45·63	56·71	1	15·51	+ 0·304	+ 0·07	..	..	..	181
97	158 52 3·50	59·91	6	15·51	+ 0·088	+ 0·00	871	398	J 49	..
98	133 25 38·49	59·98	2	15·35	+ 0·219	+ 0·07	875	406	..	..
99	72 18 7·43	59·26	4	15·33	+ 0·322	− 0·07	..	..	..	185
100	63 19 9·14	60·60	2	15·27	+ 0·338	+ 0·13	..	..	..	187
101	122 59 45·26	59·93	9	15·18	+ 0·245	− 0·12	888	415	J 50	..
102	158 12 21·47	59·91	5	15·17	+ 0·091	− 0·03	907	420	J 51	..
103	153 28 58·33	59·97	4	14·83	+ 0·131	− 0·14	937	434	..	..
104	69 53 44·18	59·71	2	14·78	+ 0·342	+ 0·03	..	..	..	193
105	69 13 20·14	58·96	14	− 14·71	+ 0·345	+ 0·02	..	..	..	194

No.	No. in B.A.C.	Magnitude	Stars' Name.	Mean R.A. 1860, Jan, 1.	Mean Year and Fraction of Year.	No. of Obs. R.A.	Annual Precess. in R.A. for 1860,	Secular Variation of Precess. in R.A.	Annual proper Motion in R.A.
				h m s	1800		s	s	s
106	937	3·4	Eridani..........θ	2 52 57·15	59·94	6	+2·280	−0·0001	−0·008
107	949	2·3	92 Ceti......... α	2 54 57·80	59·23	13	3·129	+0·0098	−0·002
108	966	6	53 Arietis...........	2 59 33·13	57·84	1	3·367	+0·0161	−0·005
109	982	5	Hydri............. θ	3 1 59·63	59·85	1	0·056	+0·0735	+0·001
110	986	4·5	57 Arietis......... δ	3 3 37·74	58·32	8	3·406	+0·0171	+0·010
111	997	3·3	12 Eridani..	3 6 7·44	59·90	2	2·522	+0·0011	+0·025
112	999	4·5	58 Arietis......... ζ	3 6 51·55	60·00	11	+3·436	+0·0177	−0·006
113	1038	5	Lacaille 1105......	3 12 25·46	59·97	1	−2·310	⊣ 0·2800	..
114	1034	5	61 Arietis........ r¹	3 13 8·71	58·73	1	+3·448	+0·0175	−0·001
115	1044	4·5	Lacaille 1060......	3 14 20·28	59·93	7	2·117	+0·0017	+0·266
116	1043	2	33 Persei.......... a	3 14 20·63	59·71	1	4·240	+0·0484	+0·002
117	1056	5	Lacaille 1092......	3 16 24·53	59·95	6	+0·639	+0·0368	+0·011
118	1070	5	Hydri ι............	3 19 31·46	59·93	4	−1·696	+0·1992	+0 040
119	..	5·6		3 25 2·39	60·00	4	+2·096	+0·0024	..
120	1109	6	Lacaille 1138.......	3 28 56·96	59·95	2	2·403	+0·0018	..
121	1125	5	Lacaille 1161......	3 32 4·33	59·93	10	2·152	+0·0024	+0·000
122	1126	6	11 Tauri............	3 32 24·88	58·08	4	3·568	+0·0189	−0·002
123	1147	4	17 Tauri............	3 36 34·11	57·60	5	3·547	+0·0179	0·000
124	1150	5	Lacaille 1191.......	3 36 40·82	59·94	6	2·384	+0·0023	−0·004
125	1159	5	Eridani............. υ¹	3 37 38·77	59·96	6	2·230	+0·0023	−0·006
126	1158	6·5	25 Eridani.........	3 37 47	..	..	3·058	+0·0078	0·000
127	1161	5	23 Tauri............	3 38 1·25	59·86	5	3·546	+0·0177	+0·003
128	1166	3	25 Tauri........... η	3 39 10·07	58·48	8	3·551	+0·0177	−0·001
129	1176	4	27 Tauri............	3 40 50·59	59·29	7	3·552	+0·0176	−0·001
130	1197	4	Lacaille 1253.......	3 42 27·27	59·94	2	0·679	+0·0294	+0·043
131	..	8		3 43 2·54	60·93	1	3·582	+0·0178	..
132	1199	5	Lacaille 1244.......	3 43 25·90	59·94	5	2·206	+0·0026	+0·009
133	..	8	Lalande 7114.......	3 43 49·39	60·92	3	3·583	+0·0178	..
134	1201	4	Eridani............ υ²	3 44 12·95	59·97	4	2·248	+0·0026	−0·008
135	1217	4	33 Eridani........ r²	3 47 45·41	59·94	2	2·549	+0·0031	+0·009
136	1220	5	Eridani............ υ³	3 48 18·87	59·90	1	2·282	+0·0026	−0·003
137	1221	6	32 Tauri............	3 48 36·04	57·84	1	+3·528	+0·0162	+0·006
138	1230	3	Hydri,............. γ	3 49 27·07	59·09	47	−1·034	+0·1080	+0·012
139	1234	3	34 Eridani....... γ¹	3 51 29·95	59·57	13	+2·792	+0·0047	+0·002
140	1257	5·4	37 Tauri......... A¹	3 56 25·42	59·72	9	+3·529	+0·0154	+0·004

No.	Mean N P.D. 1860, Jan. 1.	Mean Year and Fraction of Year.	No. of Obs. of N.P.D.	Annual Precess. in N.P.D. for 1860.	Secular Variation of Precess. in N.P.D.	Annual Proper Motion in N.P.D.	No. for reference.			
							Lacaille.	Brisbane.	Fallows or Johnson.	Greenwich or Henderson.
	° ′ ″	1800		″	″	″				
106	130 52 2·14	59·94	6	− 14·61	+ 0·234	− 0·05	950	446	J 54	H 67
107	86 27 42·95	59·13	31	14·49	+ 0·322	+ 0·11	..	453	52	197
108	72 39 46·54	57·84	1	14·21	+ 0·353	+ 0·01	..	..	..	204
109	162 26 57·52	59·85	1	14·06	+ 0·012	− 0·01	1001	482	J 58	..
110	70 48·19·53	58·84	33	13·96	+ 0·363	+ 0·00	..	..	..	208
111	119 32 27·89	59·90	2	13·80	-|- 0·272	− 0·62	1000	493	55·J 59	..
112	69 28 37·76	59·78	7	13·75	+ 0·371	+ 0·07	..	..	..	211
113	169 31 8·98	59·36	3	13·39	− 0·244	..	1105	..	..	..
114	69 21 37·03	58·73	1	13·35	+ 0·381	+ 0·03	..	..	..	218
115	133 36 27·11	59·93	7	13·27	+ 0·237	− 0·75	1060	530	J 62	..
116	40 38 23·62	59·71	1	13·27	+ 0·470	+ 0·05	..	..	..	220
117	157 26 10·27	59·95	6	13·13	+ 0·076	+ 0·15	1092	540	..	..
118	167 53 53·29	59·93	4	12·92	− 0·183	− 0·05	1131	554	..	..
119	133 6 57·26	60·00	4	12·55	+ 0·243	..	..	..	..	..
120	122 20 41·57	59·95	2	12·29	+ 0·282	+ 0·09	1138	569	..	..
121	130 44 8·67	59·93	10	12·07	+ 0·256	+ 0·04	1161	578	61·J 67	..
122	65 7 35·83	58·21	5	12·04	+ 0·421	+ 0·02	..	..	..	249
123	66 19 47·78	58·36	10	11·75	+ 0·424	+ 0·04	..	..	..	258
124	122 23 16·05	59·94	6	11·74	+ 0·287	+ 0·02	1191	589	J 69	..
125	127 45 26·49	59·96	6	11·67	+ 0·269	+ 0·09	1198	591	J 70	..
126	90 44 23·98	59·99	1	11·66	+ 0·368	+ 0·04	..	..	..	262
127	66 29 26·93	59·86	1	11·65	+ 0·426	+ 0·05	..	..	..	263
128	66 19 51·34	58·86	29	11·56	+ 0·428	+ 0·06	..	..	..	265
129	66 22 40·88	59·23	6	11·44	+ 0·430	+ 0·07	..	..	..	269
130	155 14 53·48	59·94	2	11·33	+ 0·087	− 0·03	1253	..	J 74	..
131	65 15 47·47	60·93	1	11·28	+ 0·437	..	..	..	..	..
132	128 3 0·18	59·94	5	11·26	+ 0·271	+ 0·04	1244	610	65·J 75	..
133	65 15 17·74	60·92	3	11·23	+ 0·437	..	..	..	..	..
134	126 37 35·13	59·97	4	11·20	+ 0·277	+ 0·07	1248	612	J 76	..
135	115 1 44·70	59·94	2	10·94	+ 0·316	+ 0·09	1270	618	..	494*
136	125 8 54·03	59·90	1	10·90	+ 0·284	+ 0·05	1275	620	J 78	..
137	67 55 42·76	57·84	1	10·88	+ 0·437	+ 0·14	..	..	..	278
138	164 40 1·88	59·07	52	10 82	− 0·122	− 0·12	1322	629	68·J 79	H 15
139	103 54 33·95	59·60	30	10·67	+ 0·349	+ 0·12	..	..	69·J 80	281
140	68 18 14·14	59·72	9	− 10·30	+ 0·446	+ 0·09	..	..	..	288

119. This is the nearest bright Star to the place given for B.A.C. 1088.

No.	No. in B.A.C.	Magnitude	Star's Name.	Mean R.A. 1860, Jan. 1.	Mean Year and Fraction of Year	No. of Obs. of R.A.	Annual Precess. in R.A. for 1860.	Secular Variation of Precess. in R.A.	Annual Proper Motion in R.A.
				h m s	1800				
141	1259	5	Reticuli.......... δ	3 56 32·12	59·96	4	+ 0·932	+ 0·0198	− 0·004
142	1270	5	Reticuli........... γ	3 58 52·56	59·96	3	0·849	+ 0·0215	− 0·015
143	1271	5	Reticuli........... ι	3 59 2·43	59·92	2	0·947	+ 0·0191	+ 0·010
144	1284	5·6	37 Eridani..........	4 3 33	..	..	2·923	+ 0·0058	− 0·002
145	1290	4·5	38 Eridani....... o¹	4 5 2·02	58·81	14	2·924	+ 0·0058	− 0·002
146	1299	5	Horologii.......... δ	4 6 7·73	59·97	7	2·000	+ 0·0039	+ 0·013
147	1303	5	39 Eridani........ A	4 7 44	..	..	2·851	+ 0·0051	..
148	1315	5	Horologii.......... a	4 9 21·77	59·98	5	1·982	+ 0·0041	− 0·001
149	1326	5·6	52 Tauri........ φ	4 11 45·03	57·91	1	3·679	+ 0·0166	− 0·003
150	1331	4	Doradûs........... γ	4 12 21·61	59·93	3	1·555	+ 0·0076	+ 0·004
151	1333	3·4	41 Eridani....... v⁴	4 12 35·90	60·00	3	2·263	+ 0·0031	+ 0·002
152	1336	3·4	Reticuli.......... a	4 12 37·75	60·01	3	0·748	+ 0·0216	+ 0·005
153	1341	6·5	59 Tauri........ χ¹	4 14 4·09	57·02	2	3·638	+ 0·0155	+ 0·005
154	1348	5	Lacaille 1424.......	4 14 50·86	60·01	5	1·890	+ 0·0046	− 0·002
155	1356	6	64 Tauri.......... δ²	4 16 1·62	56·71	1	3·442	+ 0·0119	+ 0·007
156	1358	5	Reticuli........... θ	4 16 6·69	60·07	1	0·651	+ 0·0232	− 0·009
157	1362	5·4	65 Tauri.......... κ¹	4 17 1·82	56·87	1	3·558	+ 0·0137	+ 0·004
158	1367	5·4	69 Tauri.......... v¹	4 17 56·09	58·91	12	3·572	+ 0·0138	+ 0·007
159	1372	4	Eridani............ v³	4 18 46·78	59·97	6	2·246	+ 0·0032	+ 0·005
160	1383	5	Reticuli............ η	4 20 23·00	60·03	4	0·616	+ 0·0231	+ 0·013
161	1376	4·3	74 Tauri......... ε	4 20 26·70	57·68	11	3·487	+ 0·0121	+ 0·005
162	1413	5	Cœli............... δ	4 26 32·88	59·99	8	+ 1·834	+ 0·0050	− 0·006
163	1426	6	Mensæ............. δ	4 27 33·65	56·63	11	− 4·307	+ 0·2777	..
164	1420	1	87 Tauri.......... a	4 27 53·41	58·75	26	+ 3·430	+ 0·0106	+ 0·004
165	1422	4	50 Eridani....... v⁶	4 28 1·11	60·06	3	2·360	+ 0·0033	− 0·007
166	1433	4·3	52 Eridani....... v⁷	4 30 6·60	59·96	1	2·334	+ 0·0033	− 0·003
167	1435	5·6	51 Eridani......... c	4 30 33	..	..	3·012	+ 0·0060	..
168	1438	3	Doradûs........... a	4 30 58·56	60·02	3	+ 1·283	+ 0·0098	+ 0·011
169	1454	5·6	Lacaille 1639........	4 33 32·33	60·16	10	− 5·653	+ 0·3730	− 0·006
170	1449	4·5	94 Tauri.......... τ	4 33 50·74	59·89	17	+ 3·592	+ 0·0122	0·000
171	1458	4·5	Cœli............... a	4 36 3·12	60·01	6	1·943	+ 0·0042	− 0·016
172	1464	5	Cœli .... β	4 37 6·48	59·99	1	2·115	+ 0·0036	− 0·006
173	1473	5	Pictoris............. λ	4 39 11·36	60·04	1	1·537	+ 0·0068	+ 0·001
174	1483	5·6	Piazzi IV. 202......	4 41 11·66	60·07	3	2·030	+ 0·0039	− 0·007
175	1506	5·6	Lacaille 1626........	4 45 43·09	60·06	6	+ 1·948	+ 0·0041	− 0·005

No.	Mean N.P.D. 1860. Jan. 1.	Mean Year and Fraction of Year.	No. of Obs. of N.P.D.	Annual Precess. in N.P.D. for 1860.	Secular Variation of Precess. in N.P.D.	Annual Proper Motion in N.P.D.	No. for reference.			
							Lacaille.	Brisbane.	Fallows or Johnson.	Greenwich or Henderson.
	° ′ ″	1800		″	″	″				
141	151 47 47·29	59·95	5	− 10·29	+ 0·121	+ 0·02	1338	642	72.J 83	..
142	152 33 3·40	59·96	3	10·11	+ 0·111	+ 0·01	1357	653	J 84	..
143	151 28 20·04	59·92	2	10·10	+ 0·124	− 0·03	1355	654	..	..
144	97 17 33·68	59·93	1	9·76	+ 0·377	+ 0·04	..	..	74	299
145	97 12 19·40	58·98	29	9·64	+ 0·378	− 0·07	..	..	75.J 85	302
146	132 21 39·25	59·97	7	9·56	+ 0·260	+ 0·00	1382	668	..	..
147	100 36 24·37	59·91	1	9·44	+ 0·371	..	..	..	J 86	..
148	132 38 28·92	59·97	6	9·31	+ 0·260	+ 0·23	1398	674	77.J 88	..
149	62 59 15·63	57·91	1	9·13	+ 0·482	+ 0·04	..	..	..	310
150	141 50 29·24	59·95	4	9·08	+ 0·206	− 0·10	1417	682	J 90	..
151	124 8 32·24	60·00	3	9·06	+ 0·299	+ 0·01	1411	681	J 89	..
152	152 49 29·94	60·01	3	9·06	+ 0·101	− 0·07	1423	683	80.J 91	H 24
153	64 42 18·09	57·02	2	8·94	+ 0·479	+ 0·04	..	..	..	312
154	134 36 18·44	60·01	5	8·88	+ 0·251	− 0·03	1424	687	..	..
155	72 53 1·86	56·71	1	8·79	+ 0·455	+ 0·04	..	..	..	315
156	153 35 45·52	60·07	1	8·78	+ 0·089	+ 0·05	1443	695	J 93	..
157	68 1 47·72	56·87	1	8·71	+ 0·471	+ 0·05	..	..	..	316
158	67 30 26·68	58·91	12	8·64	+ 0·474	+ 0·05	..	..	..	317
159	124 20 39·36	59·97	6	8·57	+ 0·300	− 0·03	1441	699	81.J 94	..
160	153 43 8·70	60·03	4	8·45	+ 0·085	− 0·17	1473	707	84.J 95	..
161	71 8 0·43	58·03	19	8·44	+ 0·465	+ 0·03	..	..	..	320
162	135 15 22·29	59·99	8	7·95	+ 0·249	+ 0·04	1512	727	85.J 96	..
163	170 32 15·09	57·12	15	7·87	− 0·574	− 0·11	1579	743	..	..
164	73 46 32·51	58·61	103	7·85	+ 0·464	+ 0·17	..	730	87	327
165	120 3 7·05	60·06	3	7·84	+ 0·320	+ 0·23	1513	732	..	328
166	120 51 5·98	59·96	1	7·67	+ 0·318	+ 0·02	1529	740	88.J 99	330
167	92 45 24·69	59·95	1	7·64	+ 0·409	..	..	..	..	..
168	145 20 8·47	60·02	3	7·60	+ 0·176	+ 0·04	1539	744	89.J 100	H 41
169	171 53 32·03	59·91	13	7·39	− 0·764	− 0·17	1639	764	..	..
170	67 18 54·81	59·99	15	7·36	+ 0·491	+ 0·02	..	..	..	336
171	132 7 58·13	60·05	6	7·18	+ 0·268	+ 0·11	1556	757	90.J 103	..
172	127 25 11·66	59·99	1	7·10	+ 0·292	− 0·20	1559	762	J 104	..
173	140 44 45·66	60·04	1	6·93	+ 0·213	− 0·02	1585	772	..	..
174	129 36 39·67	60 07	3	6·76	+ 0·282	− 0·04	1594	779	..	..
175	131 33 50·62	60·06	6	− 6·39	+ 0·273	− 0·10	1626	799	..	..

No.	No. in B.A.C.	Magnitude	Star's Name.	Mean R.A. 1860, Jan. 1.	Mean Year and Fraction of Year.	No. of Obs. of R.A.	Annual Precess. in R.A. for 1860.	Secular Variation of Precess. in R.A.	Annual Proper Motion in R.A.
				h m s	1800		s	s	s
176	1520	3	3 Aurigæ............ ι	4 47 52·80	59·02	11	+ 3·896	+ 0·0146	— 0·003
177	1519	6	Piazzi IV. 239.....	4 47 38	..	..	3·077	+ 0·0058	..
178	1528	6·5	98 Tauri......... λ	4 49 35·46	58·81	2	3·662	+ 0·0112	+ 0·001
179	1552	5·4	65 Eridani....... ψ	4 54 39	..	..	2·906	+ 0·0045	+ 0·002
180	1551	5	102 Tauri,......... ι	4 54 43·92	59·02	5	3·575	+ 0·0095	+ 0·004
181	1559	5	Lacaille 1686.......	4 56 28·27	60·00	6	+ 2·432	+ 0·0033	+ 0·005
182	1587	4·5	Lacaille 1752.......	4 59 14·60	59·07	10	— 1·796	+ 0·0702	..
183	1573	5·	Cœli............... γ¹	4 59 22·33	60·00	4	+ 2·146	+ 0·0034	+ 0·007
184	1574	5·6	Cœli .... γ²	4 59 26·35	60·06	4	2·138	+ 0·0034	0·000
185	1570	6·5	106 Tauri.......... λ	4 59 31·46	56·05	1	3·548	+ 0·0087	— 0·002
186	1575	4·3	2 Leporis.......... ε	4 59 32·13	57·97	11	2·536	+ 0·0033	0·000
187	1572	6	103 Tauri..........	4 59 34·86	58·81	2	3·649	+ 0·0097	+ 0·004
188	1579	6	66 Eridani...........	4 59 50	..	..	2·963	+ 0·0046	+ 0·000
189	1623	1	19 Orionis........ β	5 7 48·63	58·42	42	2·880	+ 0·0040	— 0·001
190	1637	6	109 Tauri,......... π	5 10 52·17	59·04	2	3·599	+ 0·0078	+ 0·001
191	1650	5	Columbæ. .... ο	5 12 26·29	60·05	3	+ 2·155	+ 0·0032	+ 0·010
192	1659	5	Doradûs............ θ	5 13 52·22	35·96	6	— 0·067	+ 0·0207	..
193	1660	5·6	22 Orionis......... ο	5 14 37·10	58·98	1	+ 3·060	+ 0·0043	..
194	1665	5·6	23 Orionis....... m	5 15 28	..	..	3·150	+ 0·0047	+ 0·001
195	1672	5	Pictoris........... ζ	5 15 56·27	60·04	3	1·465	+ 0·0055	+ 0·003
196	1681	2	112 Tauri,........ β	5 17 26·64	58·41	17	3·785	+ 0·0083	+ 0·003
197	1704	5	Pictoris........... κ	5 19 47·46	60·06	4	1·100	+ 0·0072	— 0·004
198	1712	5·6	Pictoris........... θ	5 21 35·88	40·14	2	1·358	+ 0·0056	..
199	1723	5	25 Aurigæ........ χ	5 23 37·02	58·60	6	3·900	+ 0·0081	+ 0·003
200	1730	2	34 Orionis........ δ	5 24 51·31	57·64	15	3·063	+ 0·0038	+ 0·001
201	1739	4	Columbæ.......... ε	5 26 14·63	60·03	5	2·126	+ 0·0031	+ 0·002
202	1741	3	11 Leporis....... α	5 26 33·36	60·05	1	2·644	+ 0·0029	+ 0·001
203	1765	2	46 Orionis........ ε	5 29 6·55	56·62	3	3·042	+ 0·0036	— 0·002
204	1767	3·4	123 Tauri........ ζ	5 29 16·78	59·42	4	3·582	+ 0·0055	0·000
205	1791	4	Doradûs............ β	5 32 24·82	60·04	5	0·513	+ 0·0091	— 0·003
206	1802	2	Columbæ,......... α	5 34 34·81	57·78	21	+ 2·171	+ 0·0028	+ 0·008
207	1819	5·6	Mensæ............. γ	5 37 27·00	60·07	1	— 2·445	+ 0·0372	..
208	1841	5·6	Columbæ.......... μ	5 40 47·75	60·07	5	+ 2·228	+ 0·0025	— 0·005
209	1855	5	Lacaille 2003........	5 42 34·69	60·07	1	1·660	+ 0·0033	— 0·001
210	1861	4·5	Pictoris............. β	5 43 58·19	60·03	1	+ 1·418	+ 0·0038	..

192. The R.A. has been brought up with precession alone from 1835.
198. The R.A. has been brought up with precession alone from 1840.

No.	Mean N.P.D. 1860, Jan. 1.	Mean Year and Fraction of Year.	No. of Obs. of N.P.D.	Annual Precess. in N.P.D. for 1860.	Secular Variation of Precess.in N.P.D.	Annual Proper Motion in N.P.D.	No. for reference.			
							Lacaille.	Brisbane.	Fallows or Johnson.	Greenwich or Henderson.
	° ′ ″	1800		″	″	″				
176	57 3 34·43	58·95	14	− 6·21	+ 0·543	+ 0·02	..	..	..	353
177	89 45 47·07	60·03	1	6·23	+ 0·429	..	..	..	..	..
178	65 10 10·23	58·81	2	6·07	+ 0·512	+ 0·06	..	..	..	625*
179	97 22 56·43	60·00	4	5·64	+ 0·409	− 0·01	..	..	J 108	364
180	68 36 49·56	58·20	5	5·64	+ 0·503	+ 0·06	..	..	..	365
181	116 28 34·58	60·00	6	5·49	+ 0·343	+ 0·10	1686	846	..	..
182	165 9 0·51	59·07	17	5·26	− 0·251	− 0·08	1752	872	..	..
183	125 40 36·71	60·00	4	5·24	+ 0·304	+ 0·09	1712	858	J 110	..
184	125 54 7·18	60·06	4	5·24	+ 0·303	− 0·10	1713	860	..	..
185	69 46 9·87	56·05	1	5·23	+ 0·502	+ 0·04	..	..	..	369
186	112 33 42·22	58·82	26	5·23	+ 0·359	+ 0·07	..	..	J 109	370
187	65 55 24·63	58·81	2	5·23	+ 0·516	− 0·05	..	..	..	372
188	94 50 46·36	59·95	1	5·20	+ 0·420	+ 0·04	..	..	..	373
189	98 21 59·13	57·88	109	4·53	+ 0·411	+ 0·02	..	893	99.J 116	383
190	68 3 7·25	59·04	2	4·27	+ 0·515	− 0·03	..	..	..	386
191	125 2 3·65	60·05	3	4·13	+ 0·310	+ 0·31	1793	914	..	..
192	157 20 34·68	60·03	1	4·01	− 0·008	− 0·04	1828	922	..	..
193	90 31 26·85	58·98	1	3·95	+ 0·439	..	..	..	..	..
194	86 35 37·60	59·95	1	3·87	+ 0·453	+ 0·02	..	..	..	391
195	140 45 29·20	60·02	5	3·83	+ 0·212	− 0·14	1825	930	..	..
196	61 30 53·49	58·51	35	3·70	+ 0·544	+ 0·20	..	932	..	395
197	146 16 1·24	60·06	4	3·50	+ 0·160	− 0·09	1853	956	..	..
198	142 26 23·60	60·06	1	3·34	+ 0·197	+ 0·04	1863	962	..	..
199	57 54 56·73	58·34	5	3·17	+ 0·563	− 0·02	..	..	..	407
200	90 24 22·00	57·61	35	3·06	+ 0·443	+ 0·04	..	968	105.J 122	409
201	125 34 31·74	60·03	5	2·94	+ 0·308	+ 0·07	1883	970	J 124	..
202	107 55 31·53	58·94	4	2·92	+ 0·383	+ 0·00	..	..	107.J 125	413
203	91 17 40·75	56·56	5	2·69	+ 0·441	+ 0·01	..	..	110.J 128	423
204	68 56 48·21	60·04	9	2·68	+ 0·519	+ 0·05	..	..	109	425
205	152 34 54·54	60·03	7	2·41	+ 0·075	− 0·06	1948	1003	J 131	..
206	124 9 3·27	57·97	39	2·22	+ 0·316	0·00	1938	1010	114.J 133	429
207	166 26 17·00	59·23	7	1·97	− 0·354	− 0·31	2027	1032	..	..
208	122 21 43·60	60·07	5	1·68	+ 0·325	+ 0·04	1982	1035	116.J 136	..
209	136 39 1·52	60·07	1	1·52	+ 0·242	+ 0·00	2003	1043	..	..
210	141 7 8·30	60·07	3	− 1·40	+ 0·207	− 0·06	2021	1051	..	..

No.	No. in B.A.C.	Magnitude.	Star's Name.	Mean R.A. 1860, Jan. 1.	Mean Year and Fraction of Year.	No. of Obs. R.A.	Annual Precess. in R.A. for 1860.	Secular Variation of Precess. in R.A.	Annual Proper Motion in R.A.
				h m s	1800		s	s	s
211	1870	5·6	Mensæ............ ι	5 44 11·88	73·04	3	— 3·721	+ 0·0446	..
212	1868	4·5	Doradûs............ δ	5 44 31·57	60·05	2	+ 0·105	+ 0·0082	— 0·002
213	1863	5	136 Tauri.........	5 44 31·68	57·95	5	3·769	+ 0·0040	— 0·001
214	1878	3	Columbæ......... β	5 46 1·51	60·03	5	2·109	+ 0·0026	+ 0·002
215	1876	5·4	54 Orionis....... χ¹	5 46 5·66	60·09	1	3·564	+ 0·0034	— 0·016
216	1883	Var.	58 Orionis........ α	5 47 35·55	57·25	28	3·245	+ 0·0028	+ 0·001
217	1890	5	Lacaille 2052.......	5 47 43·23	59·99	1	1·354	+ 0·0034	..
218	1891	5	Columbæ......... λ	5 48 1·99	60·02	1	+ 2·177	+ 0·0026	0·000
219	1898	5·6	Lacaille 2138......	5 48 23·27	73·04	3	— 4·976	+ 0·0450	..
220	1896	5·6	139 Tauri.........	5 49 18·45	58·44	2	+ 3·722	+ 0·0032	— 0·001
221	1905	5	Doradûs............ ε	5 50 2·47	60·07	1	— 0·065	+ 0·0069	— 0·003
222	1922	4	Columbæ......... γ	5 52 34·37	60·06	4	+ 2·126	+ 0·0025	— 0·005
223	1933	5	Lacaille 2099......	5 54 51·72	60·05	3	1·833	+ 0·0026	+ 0·001
224	1958	5·4	67 Orionis........ ν	5 59 34·68	57·81	8	3·425	+ 0·0018	+ 0·001
225	1964	5·6	Lacaille 2137......	6 0 26·68	60·08	9	1·733	+ 0·0025	— 0·009
226	1971	6	3 Geminorum......	6 1 13·85	60·98	1	3·643	+ 0·0015	+ 0·006
227	1982	5	Columbæ.......... θ	6 2 43·60	60·07	6	2·056	+ 0·0024	— 0·007
228	1981	6	5 Geminorum......	6 2 57·10	59·12	2	3·680	+ 0·0012	0·000
229	2002	3·4	7 Geminorum.... η	6 6 25·58	59·60	7	3·627	+ 0·0008	— 0·007
230	2001	5·4	44 Aurigæ......... κ	6 6 27·47	58·34	3	3·830	+ 0·0005	— 0·005
231	2013	5·6	Lacaille 2201.......	6 7 34·26	60·08	9	1·168	+ 0·0020	— 0·006
232	2034	4·5	Columbæ.......... κ	6 11 34·32	60·08	5	2·134	+ 0·0021	+ 0·001
233	2047	3	13 Geminorum... μ	6 14 29·42	59·26	11	3·627	— 0·0002	+ 0·005
234	2051	3·2	1 Canis Majoris.. ζ	6 14 56·39	60·06	10	+ 2·302	+ 0·0019	+ 0·002
235	2085	6	Mensæ.	6 16 37·44	56·21	6	—15·624	— 0·3305	— 0·003
236	2082	6·5	48 Aurigæ	6 19 34·07	58·60	4	+ 3·859	— 0·0018	+ 0·002
237	2090	5·4	18 Geminorum... ν	6 20 39·08	60·09	1	3·565	— 0·0008	— 0·002
238	2096	1	Argûs.............. α	6 20 50·54	59·81	19	1·329	+ 0·0009	— 0·002
239	2109	4·5	Lacaille 2295.......	6 22 58·80	60·09	6	+ 2·225	+ 0·0018	+ 0·001
240	2119	5·6	Doradûs.......... π¹	6 23 57	..	..	— 0·563	— 0·0096	..
241	2137	5	Puppis............. Z	6 26 22·48	60·08	3	+ 1·481	+ 0·0010	— 0·010
242	2145	5·6	Doradûs.......... π²	6 26 39·99	56·13	5	— 0·501	— 0·0104	..
243	2163	2·3	24 Geminorum... γ	6 29 37·41	57·89	9	+ 3·465	— 0·0014	+ 0·001
244	2170	6	54 Aurigæ..	6 30 43·20	57·02	1	3·788	— 0·0033	+ 0·001
245	2176	5	Lacaille 2383........	6 31 53·48	60·07	6	+ 1·324	+ 0·0002	0·000

219. The large proper Motion of this Star is confirmed by the result from three Observations on, 1873, January 8, 15, and 22. The mean place for 1873, January 1, given by these Observations is,

R.A. 5 47 18·62. .   N.P.D. 170 33 41·08.

No,	Mean N.P.D. 1860. Jan. 1.	Mean Year and Fraction of Year.	No. of Obs. of N.P.D.	Annual Precess. in N.P.D. for 1860.	Secular Variation of Precess. in N.P.D	Annual Proper Motion in N.P.D.	No. for reference.			
							Lacaille.	Brisbane.	Fallows or Johnson.	Greenwich or Henderson
	° ′ ″	1800		″	″	″				
211	168 53 23·01	59·07	5	− 1·37	− 0·540	− 0·08	2097	1068	..	..
212	155 47 17·33	60·05	2	1·35	+ 0·016	+ 0·02	2045	1060	118.J 139	..
213	62 25 29·37	57·55	6	1·35	+ 0·549	+ 0·07	..	..	..	439
214	125 49 23·46	60·03	5	1·22	+ 0·308	− 0·28	2029	1063	119.J 140	441
215	69 45 11·42	60·09	1	1·22	+ 0·519	+ 0·10	..	..	..	442
216	82 37 21·20	57·08	90	1·09	+ 0·473	0·00	..	1064	120	444
217	142 8 31·81	59·99	1	1·07	+ 0·198	+ 0·07	2052	1074	..	..
218	123 50 5·11	60·02	1	1·05	+ 0·318	− 0·09	2044	1073	..	..
219	170 34 7·97	59·07	4	1·02	− 0·724	− 1·10	2138	1096	..	..
220	64 4 2·79	58·07	2	0·94	+ 0·543	+ 0·01	..	..	..	446
221	156 56 9·42	60·07	1	0·87	− 0·009	− 0·06	2093	1091	J 142	..
222	125 18 1·78	60·06	4	0·65	+ 0·310	+ 0·01	2084	1097	121.J 143	..
223	132 49 29·03	60·05	3	0·45	+ 0·268	+ 0·02	2099	1107	J 144	..
224	75 13 6·68	58·47	14	− 0·04	+ 0·500	+ 0·02	..	..	..	457
225	135 2 15·84	60·07	11	+ 0·04	+ 0·253	− 0·22	2137	1131	..	..
226	66 52 5·36	60·98	1	0·11	+ 0·531	+ 0·05	..	1127	..	462
227	127 14 7·50	60·07	6	0·24	+ 0·300	− 0·01	2153	1145	J 146	..
228	65 33 10·05	59·12	2	0 26	+ 0·537	+ 0·07	..	..	..	465
229	67 27 22·66	59·81	11	0·56	+ 0·529	+ 0·02	..	1166	..	467
230	60 27 16·29	58 05	10	0·56	+ 0·558	+ 0·29	..	..	..	468
231	144 56 18·91	60·08	9	0·66	+ 0·170	+ 0·02	2201	1177	..	..
232	125 5 46·13	60·08	5	1·01	+ 0·310	+ 0·03	2213	1191	J 148	473
233	67 25 7·06	59·46	22	1·27	+ 0·527	+ 0·14	..	1202	..	477
234	120 0 13·31	60·06	10	1·31	+ 0·334	− 0·02	2229	1207	126.J 149	478
235	175 55 12·70	56·21	6	1·45	− 2·273	+ 0·04	2512	1269	..	..
236	59 25 28·06	58·59	4	1·71	+ 0·560	+ 0·04	..	..	..	486
237	69 42 10·18	60·09	2	1·81	+ 0·517	+ 0·01	..	1235	..	488
238	142 37 13·04	59·41	24	1·82	+ 0·192	− 0·03	2291	1241	128.J 152	H 46
239	122 29 39·77	60·09	6	2·01	+ 0·322	− 0·08	2295	1147	J 153	491
240	159 54 20·73	59·07	6	2 09	− 0·082	− 0·08	2340	1259	..	..
241	140 8 30·67	60·08	3	2·30	+ 0·214	− 0·02	2333	1267	..	..
242	159 36 33·72	56·13	5	2·33	− 0·074	− 0·43	2368	1275	..	..
243	73 29 6·02	58·82	17	2·59	+ 0·500	+ 0·04	..	1280	130	501
244	61 37 5·27	57·02	1	2·68	+ 0·546	+ 0·04	..	..	..	507
245	142 51 45·47	60·06	7	+ 2·78	+ 0·190	+ 0·01	2383	1302	..	..

No.	No. in B.A.C.	Magnitude	Star's Name.	Mean R.A. 1860, Jan. 1.	Mean year and Fraction of year.	No. of Obs. of R.A.	Annual Precess. in R.A. for 1860.	Secular Variation of Precess. in R.A.	Annual Proper Motion in R.A.
				h m s	1800		•	•	s
246	2188	3	Argûs .... ν	6 33 28·60	60·05	10	+ 1·835	+ 0·0014	— 0·004
247	2194	3·4	27 Geminorum... ε	6 35 18·91	56·43	2	3·696	— 0·0034	0·000
248	2197	6	28 Geminorum......	6 35 52·95	57·02	1	3·807	— 0·0043	+ 0·003
249	2213	1	9 Canis Majoris. α	6 38 58·66	58·29	71	2·681	+ 0·0010	— 0·035
250	2231	5	Puppis .... χ	6 42 34·08	60·07	5	2·054	+ 0·0015	— 0·001
251	2233	6	36 Geminorum.. d	6 43 9·38	57·17	1	3·601	— 0·0038	+ 0·007
252	2246	4	13 Canis Majoris κ	6 44 36·74	60·08	6	2·241	+ 0·0015	— 0·002
253	2252	5	Lacaille 2486........	6 45 46·81	60·11	3	2·181	+ 0·0014	— 0·001
254	2260	4	Pictoris............ α	6 46 45·32	60·12	1	0·631	— 0·0063	— 0·005
255	2259	5	Carinæ............ B	6 46 48·76	38·21	5	1·305	— 0·0012	••
256	2273	5	18 Canis Majoris μ	6 49 (41·73)	••	••	+ 2·750	+ 0·0006	0·000
257	2290	5·6	Mensæ............ ζ	6 51 37·88	56·11	5	— 4·852	— 0·1662	••
258	2293	2·1	21 Canis Majoris ε	6 53 7·39	59·39	28	+ 2·357	+ 0·0013	0·000
259	2295	5	Puppis............ ι	6 53 17·59	60·12	4	2·197	+ 0·0014	— 0·004
260	2305	4	43 Geminorum... ζ	6 55 (48·27)	••	••	3·564	— 0·0050	— 0·001
261	2309	5·4	22 Canis Majoris...	6 56 8·52	60·02	3	2·390	+ 0·0013	— 0·003
262	2319	4·5	23 Canis Majoris γ	6 57 25·50	58·26	6	2·715	+ 0·0005	+ 0·002
263	2327	5	Puppis............ C	6 59 36·47	60·07	5	1·903	+ 0·0008	— 0·007
264	2340	5·4	46 Geminorum... r	7 2 13·61	60·02	1	3·830	— 0·0088	— 0·003
265	2345	2	25 Canis Majoris δ	7 2 41·90	60·05	2	2·439	+ 0·0011	0·000
266	2343	6	47 Geminorum......	7 2 41·94	60·15	6	3·730	— 0·0077	~ 0·003
267	2350	6	48 Geminorum......	7 3 55·87	60·99	2	3·654	— 0·0069	+ 0·004
268	2355	5	Puppis............ A	7 4 8·93	60·05	1	2·015	+ 0·0011	— 0·008
269	2374	6	53 Geminorum......	7 7 12·52	59·12	1	3·757	— 0·0087	+ 0·002
270	2380–	5	Puppis............ E	7 7 37·53	60·04	2	1·989	+ 0·0010	— 0·010
271	2410	3·4	55 Geminorum... δ	7 11 45·53	59·51	18	3·592	— 0·0072	0·000
272	2414	3	Argûs............ π	7 12 11·91	60·13	2	2·119	+ 0·0011	— 0·004
273	2427	5	Puppis............ F	7 13 46·89	60·04	2	+ 2·047	+ 0·0010	— 0·018
274	2447	5	Volantis............ δ	7 16 53·39	56·82	6	— 0·006	— 0·0249	— 0·004
275	2442	4	60 Geminorum... ι	7 17 1·62	57·80	4	+ 3·745	— 0·0100	— 0·008
276	2458	2·7	31 Canis Majoris η	7 18 33·46	60·16	5	2·373	+ 0·0011	— 0·004
277	2467	5	64 Geminorum.. b¹	7 20 36·72	58·97	1	3·751	— 0·0106	+ 0·002
278	2482	4	Argûs............ σ	7 24 47·68	60·24	1	1·909	+ 0·0005	+ 0·005
279	2484	5	Lacaille 2834.......	7 25 16·15	60·20	1	2·333	+ 0·0012	••
280	2485	2·1	66 Geminorum., α²	7 25 39·66	60·19	2	+ 3·856	— 0·0132	— 0·013

255. The R.A. has been brought up with Precession alone.

No.	Mean N P.D. 1860, Jan. 1.	Mean Year and Fraction of Year	No. of Obs. of N.P.D.	Annual Precess. in N.P.D. for 1860.	Secular Variation of Precess. in N.P.D.	Annual Proper Motion in N.P.D.	No. for reference.			
							Lacaille.	Brisbane.	Fallows or Johnson.	Greenwich or Henderson.
	° ′ ″	1800		″	″	″				
246	133 4 30·20	60·05	10	+ 2·92	+ 0·264	+ 0·01	2386	1310	132.J 156	H 60
247	64 44 3·91	57·60	11	3·08	+ 0·531	+ 0·02	..	1316	..	511
248	60 53 33·59	57·02	1	3·13	+ 0·547	+ 0·03	..	..	..	512
249	106 31 36·54	58·53	68	3·40	+ 0·384	+ 1·24	..	1337	133.J 157	517
250	127 46 37·81	60·07	5	3·70	+ 0·292	+ 0·06	2455	1359	..	..
251	68 4 40·30	57·17	1	3·76	+ 0·514	+ 0·02	..	..	..	521
252	122 20 56·82	60·08	6	3·88	+ 0·319	+ 0·03	2474	1371	135.J 158	523
253	124 12 14·33	60·11	3	3·98	+ 0·310	− 0·04	2486	1378	..	..
254	151 47 30·16	60·12	1	4·06	+ 0·088	− 0·18	2525	1389	J 161	..
255	143 27 36·76	60·10	1	4·07	+ 0·184	− 0·02	2511	1388	..	..
256	103 51 55·83	60·01	1	4·32	+ 0·390	+ 0·01	..	..	..	531
257	170 39 38·00	56·11	5	4·48	− 0·692	− 0·12	2648	1435	..	..
258	118 47 2·17	58·38	70	4·61	+ 0·333	+ 0·02	2550	1419	138.J 164	534
259	123 55 26·91	60·12	4	4·62	+ 0·310	− 0·07	2554	1421	..	..
260	69 13 41·52	56·12	3	4·84	+ 0·503	+ 0·01	..	1431	..	537
261	117 44 13·24	60·02	3	4·86	+ 0·336	+ 0·01	2581	1437	139.J 165	538
262	105 25 45·11	58·87	9	4·97	+ 0·381	+ 0·01	..	..	140.J 167	542
263	132 7 56·10	60·08	4	5·16	+ 0·266	− 0·11	2607	1462	..	..
264	59 31 43·49	60·02	1	5·38	+ 0·536	+ 0·05	..	..	..	549
265	116 10 24·52	60·05	2	5·42	+ 0·340	− 0·01	2633	1478	143.J 168	550
266	62 55 1·72	60·15	6	5·42	+ 0·521	+ 0·03	..	..	..	898*
267	65 38 27·88	60·99	2	5·52	+ 0·510	+ 0·01	..	..	..	552
268	129 25 58·51	60·05	1	5·54	+ 0·280	+ 0·05	2649	1486	J 169	..
269	61 51 49·00	59·12	1	5·80	+ 0·522	0·00	..	..	..	556
270	130 15 50·87	60·04	2	5·83	+ 0·275	+ 0·05	2672	1504	..	..
271	67 45 48·89	59·40	31	6·18	+ 0·496	+ 0·02	..	..	..	566
272	126 50 54·18	60·13	2	6·21	+ 0·291	+ 0·02	2720	1536	145.J 175	567
273	128 57 22·97	60·04	2	6·35	+ 0·280	0·00	2739	1552	..	..
274	157 42 2·97	56·82	6	6·60	− 0·004	0·00	2809	1586	J 176	..
275	61 55 39·05	57·83	12	6·62	+ 0·513	+ 0·09	..	..	..	570
276	119 1 56·61	60·16	5	6·74	+ 0·323	− 0.01	2777	1591	146.J 177	R 791
277	61 35 49·09	58·97	1	6·91	+ 0·511	+ 0·04	..	..	..	574
278	133 1 14·20	60·24	1	7·25	+ 0·257	− 0·09	2837	1631	147.J 178	..
279	120 40 15·04	60·20	1	7·29	+ 0·314	− 0·04	2834	1634	..	..
280	57 48 30·64	58·80	7	+ 7·32	+ 0·520	+ 0·08	..	1630	..	580

No.	No. in B.A.C.	Magnitude	Star's Name.	Mean R.A. 1860, Jan. 1.	Mean Year and Fraction of Year.	No. of Obs. of R.A.	Annual Precess. in R.A. for 1860.	Secular Variation of Precess. in R.A.	Annual Proper Motion in R.A.
				h m s	1800		s	s	s
281	2493	4·5	69 Geminorum... υ	7 27 17·50	59·42	10	+ 3·710	− 0·0109	− 0·001
282	2522	1	10 Canis Minoris a	7 31 58·31	59·61	30	3·192	− 0·0041	− 0·048
283	2551	4·3	77 Geminorum.. κ	7 35 59·50	59·85	8	3·635	− 0·0108	− 0·005
284	2555	1·2	78 Geminorum.. β	7 36 44·62	59·18	3	3·730	− 0·0127	− 0·049
285	2562	·5	3 Puppis............	7 38 11·32	60·16	2	2·408	+ 0·0010	− 0·002
286	2565	6	B.F. 1089...........	7 38 41·28	57·21	3	2·522	+ 0·0008	..
287	2569	7	2 Puppis............	7 39 2·69	57·24	1	2·761	− 0·0004	+ 0·002
288	2580	5	Puppis............ c	7 40 16·05	60·16	3	2·138	+ 0·0011	0·000
289	2594	5	Puppis............ o	7 42 16·08	60·21	2	2·494	+ 0·0009	− 0·004
290	2599	6·7	Lacaille 2990......	7 43 8·74	60·14	1	2·522	+ 0·0008	..
291	2602	4·3	7 Navis............ ξ	7 43 24·28	60·27	2	2·523	+ 0·0009	+ 0·001
292	2617	5	83 Geminorum.. φ	7 44 55·50	57·85	6	3·686	− 0·0130	− 0·004
293	2620	4·5	Puppis............ P	7 44 58·38	60·18	2	1·829	0·0000	− 0·002
294	2622	6	9 Puppis............	7 45 17·29	60·09	1	2·784	− 0·0006	− 0·003
295	2624	6	10 Puppis............	7 45 52·46	57·21	3	2·763	− 0·0004	+ 0·002
296	2629	5	Lacaille 3035........	7 47 1·97	60·21	3	2·256	+ 0·0012	− 0·019
297	2642	5	Lacaille 3069........	7 49 8	..	..	1·693	− 0·0011	..
298	2644	4	Puppis............ R	7 49 11·40	60·17	3	1·765	− 0·0005	− 0·002
299	2655	6	Lacaille 3081.......	7 52 5·33	60·17	3	2·392	+ 0·0013	+ 0·017
300	2660	6·5	27 Monocerotis.....	7 52 44·52	57·23	1	3·004	− 0·0027	− 0·005
301	2665	4	Argûs............ χ	7 53 13·23	60·25	2	1·532	− 0·0029	+ 0·001
302	2666	5	B.F. 1129..........	7 53 35·73	58·70	6	2·690	+ 0·0001	0·000
303	2670	5	Lacaille 3105........	7 54 12·91	40·11	2	1·727	− 0·0009	..
304	2672	5	6 Cancri............	7 54 54·80	58·73	2	3·700	− 0·0147	− 0·005
305	2710	2·3	Argûs............ ζ	7 58 39·85	60·23	2	2·111	+ 0·0013	− 0·004
306	..	7·8	Lalande 15898......	8 1 14·94	56·95	1	3·634	− 0·0140	..
307	2725	5·6	29 Monocerotis.....	8 1 33·51	57·24	2	3·020	− 0·0031	..
308	2728	3	15 Navis............	8 1 34·92	59·68	11	2·561	+ 0·0009	− 0·007
309	2730	6	14 Cancri........ ψ²	8 2 0·92	58·19	2	3·632	− 0·0141	− 0·006
310	2736	5	16 Puppis............	8 2 46·86	57·21	2	2·680	+ 0·0002	+ 0·003
311	2755	2	Argûs............ γ	8 5 13·14	60·21	3	1·850	+ 0·0001	+ 0·002
312	2769	5	20 Puppis............	8 6 53·91	59·21	3	2·759	− 0·0003	+ 0·003
313	2773	5	Volantis.......... ε	8 7 27·60	60·11	1	0·232	− 0·0362	− 0·015
314	2774	5	Puppis............. r	8 8 12·40	60·11	2	2·264	+ 0·0017	− 0·004
315	2778	4·3	17 Cancri........ β	8 8 55·22	57·24	2	+ 3·264	− 0·0071	− 0·004

303. The R.A. has been brought up with Precession alone.

No.	Mean N.P.D. 1860, Jan. 1..	Mean Year and Fraction of Year.	No. of Obs. of N.P.D.	Annual Precess. in N.P.D. for 1860.	Secular Variation of Precess.in N.P.D.	Annual Proper Motion in N.P.D.	No. for reference.			
							Lacaille.	Brisbane.	Fallows or Johnson.	Greenwich or Henderson
	° ′ ″	1800		″	″	″				
281	62 47 48·10	59·38	8	+ 7·46	+ 0·499	+ 0·11	..	..	..	581
282	84 25 10·21	59·11	61	7·83	+ 0·425	+ 1·08	..	1666	150	585
283	65 16 11·67	59·69	7	8·16	+ 0·481	+ 0·05	..	..	..	590
284	61 38 21·56	58·55	22	8·22	+ 0·493	+ 0·06	..	1704	..	592
285	118 37 20·73	60·16	2	8·33	+ 0·316	+ 0·05	2938	1717	J 180	..
286	114 20 23·54	57·21	3	8·37	+ 0·330	..	..	..	..	..
287	104 20 56·50	57·24	1	8·40	+ 0·362	0·00	..	..	..	971*
288	127 37 50·46	60·16	1	8·50	+ 0·279	0·00	2958	1735	J 181	..
289	115 35 31·82	60·21	2	8·66	+ 0·324	0·00	2981	1750	..	..
290	114 33 51·89	60·14	1	8·73	+ 0·327	− 0·28	2990	1760	..	595
291	114 30 38·39	60·27	2	8·75	+ 0·328	− 0·02	2994	1763	J 182	597
292	62 52 30·96	57·61	7	8·87	+ 0·479	+ 0·05	..	..	..	599
293	136 1 20·27	60·18	2	8·87	+ 0·235	0·00	3022	1778	152. J 184	..
294	103 31 46·38	60·09	1	8·89	+ 0·360	+ 0·33	..	..	J 183	600
295	104 29 20·12	57·21	3	8·94	+ 0·357	+ 0·05	..	..	..	985*
296	124 21 19·96	60·21	3	9·03	+ 0·290	− 0·32	3035	1797	..	..
297	119 15 1·46	60·05	1	9·19	+ 0·215	− 0·02	3069	1813	..	..
298	137 44 21·03	60·17	3	9·20	+ 0·225	− 0·07	3068	1812	J 187	..
299	119 57 36·97	60·17	3	9·42	+ 0·304	+ 0·11	3081	1825	..	607
300	93 18 4·54	57·23	1	9·47	+ 0·382	− 0·03	..	..	..	992*
301	142 36 30·71	60·25	2	9·51	+ 0·192	+ 0·02	3102	1835	156. J 188	H 47
302	108 1 3·14	59·07	8	9·54	+ 0·341	0·00	..	..	..	..
303	138 52 0·19	60·10	1	9·59	+ 0·217	+ 0·02	3105	1839	..	..
304	61 48 59·11	58·83	11	9·64	+ 0·469	+ 0·07	..	..	..	613
305	129 36 37·88	60·23	3	9·93	+ 0·263	− 0·03	3136	1876	157. J 189	H 70
306	64 2 2·91	56·95	1	10·12	+ 0·454	..	..	..	..	..
307	92 34 44·27	58·66	4	10·15	+ 0·376	..	..	..	..	..
308	113 54 11·03	57·86	49	10·15	+ 0·318	− 0·06	3153	1892	J 190	619
309	64 4 15·95	57·93	16	10·18	+ 0·452	+ 0·35	..	..	..	620
310	108 50 15·20	57·21	3	10·24	+ 0·332	− 0·03	..	..	..	621
311	136 55 32·81	60·19	2	10·42	+ 0·226	+ 0·04	3185	1917	158. J 192	H 55
312	105 22 7·41	59·21	3	10·55	+ 0·338	+ 0·03	..	..	J 193	628
313	158 12 21·01	60·11	1	10·59	+ 0·024	− 0·06	3242	1940	J 194	..
314	125 28 42·03	60·11	2	10·64	+ 0·275	+ 0·02	3212	1938	..	..
315	80 23 9·95	57·24	2	+ 10·70	+ 0·398	+ 0·05	..	..	..	630

No.	No. in B.A.C.	Magnitude	Star's Name.	Mean R.A. 1860, Jan. 1.	Mean Year and Fraction of Year.	No. of Obs. R.A.	Annual Precess. in R.A. for 1860.	Secular Variation of Precess. in R.A.	Annual Proper Motion in R.A.
				h m s	1800		s	s	s
316	2785	6	21 Puppis............	8 10 58·19	57·21	3	+2·753	—0·0002	+0·002
317	2786	6	18 Cancri......... χ	8 11 33·38	57·10	2	3·661	—0·0161	+0·005
318	2789	6	19 Cancri........ λ	8 12 12·50	59·12	2	3·582	—0·0141	—0·004
319	2795	5	Puppis............. g	8 13 18·83	60·24	7	2·254	+0·0018	—0·017
320	2802	5	Puppis............. w	8 15 52·29	60·16	5	2·363	—0·0019	+0·016
321	2807	6	22 Puppis............	8 16 12·06	57·30	1	2·824	—0·0009	0·000
322	2823	5	Velorum.......... B	8 18 13·15	60·22	1	1·847	+0·0002	—0·007
323	2832	2	Argûs............. ε	8 19 37·93	60·25	1	+1·243	—0·0091	—0·014
324	2878	7	Octantis......... A	8 20 17·23	61·32	7	—37·537	—16·1887	—0·014
325	2849	4·5	Chamæleontis.... a	8 22 4·85	60·30	1	—1·453	—0·1452	+0·024
326	2856	5	Volantis........... η	8 23 17·89	60·22	3	—0·462	—0·0770	—0·022
327	2853	·6	31 Cancri......... θ	8 23 36·58	56·73	3	+3·436	—0·0117	—0·006
328	2863	5	Volantis .... β	8 24 12·04	60·24	2	0·681	—0·0249	—0·009
329	2862	6	33 Cancri......... η	8 24 36·46	58·48	7	+3·485	—0·0130	—0·005
330	2870	5	Chamæleontis.... θ	8 24 46·31	58·84	6	—1·613	—0·1593	—0·045
331	2917	6	39 Cancri............	8 32 2·91	60·92	3	+3·466	—0·0132	—0·009
332	2918	6	40 Cancri............	8 32 8·08	58·53	3	3·465	—0·0132	—0·006
333	2926	5	Velorum........... ε	8 32 43·31	60·22	9	2·109	+0·0024	..
334	2929	·6	6 Hydræ............	8 33 23·57	57·30	1	2·849	—0·0010	—0·002
335	2935	·5	Mali................ b	8 34 37·33	60·20	2	2·346	+0·0027	+0·018
336	2937	4·5	43 Cancri........ γ	8 35 10·76	56·69	2	3·492	—0·0142	—0·011
337	2947	·5	Velorum.......... b	8 35 58·90	60·28	1	1·990	+0·0018	—0·006
338	2950	·4	Argûs............. o	8 36 17·15	60·24	1	1·723	—0·0009	+0·001
339	2953	4	47 Cancri......... δ	8 36 43·53	59·54	13	3·422	—0·0124	—0·002
340	2962	5	Carinæ............. d	8 37 (31·53)	..	..	1·334	—0·0079	..
341	2971	3·4	11 Hydræ......... ε	8 39 21·56	59·51	8	3·197	—0·0071	—0·013
342	2976	·6	Piazzi VIII. 167...	8 40 9·18	57·22	2	3·047	—0·0041	..
343	2979	3	Argûs............. δ	8 40 50·02	60·21	1	1·656	—0·0018	—0·005
344	2981	·5	Velorum.......... a	8 41 16·87	60·18	2	2·034	+0·0023	—0·009
345	2987	6	14 Hydræ............	8 42 19·70	57·24	2	3·020	—0·0035	—0·001
346	2990	7	Piazzi VIII. 179...	8 42 44·27	57·30	3	3·412	—0·0125	..
347	2998	5	Carinæ .... f	8 43 5·07	60·23	6	1·556	—0·0034	..
348	3011	6	15 Hydræ	8 44 42	..	..	+2·955	—0·0024	—0·002
349	3023	·5	Chamæleontis..... η	8 45 59·06	60·21	2	—1·829	—0·2147	—0·013
350	3035	6	60 Cancri............	8 48 16·72	58·08	2	+3·286	—0·0096	—0·002

340. The R.A. has been brought up from Johnson.

No.	Mean N.P.D. 1860, Jan. 1.	Mean Year and Fraction of Year.	No. of Obs. of N.P.D.	Annual Precess. in N.P.D. for 1860.	Secular Variation of Precess. in N.P.D.	Annual Proper Motion in N.P.D.	No. for reference.			
							Lacaille.	Brisbane.	Fallows or Johnson.	Greenwich or Henderson.
	o ′ ″	1800		′	′	′				
316	105 51 14·23	57·21	3	+ 10·85	+ 0·333	— 0·01	..	..	..	1028*
317	62 19 55·63	57·10	2	10·89	+ 0·444	+ 0·37	..	..	..	632
318	65 32 24·41	59·12	2	10·94	+ 0·433	+ 0·04	..	..	..	633
319	126 13 38·12	60·24	7	11·02	+ 0·270	— 0·11	3259	1968	159.J 195	..
320	122 36 40·19	60·16	5	11·21	+ 0·281	+ 0·08	3277	1979	..	639
321	102 36 27·85	57·30	1	11·23	+ 0·336	— 0·01	..	..	..	1034*
322	138 2 33·08	60·20	4	11·38	+ 0·217	— 0·01	3308	2003	..	..
323	149 3 35·32	60·25	1	11·48	+ 0·142	— 0·03	3327	2012	160.J 196	H 32
324	178 27 23·32	57·75	24	11·52	— 4·486	0·00	..	2298	..	H 5
325	166 28 33·00	60·30	1	11·65	— 0·178	— 0·12	3400	2048	J 197	..
326	162 56 47·30	60·22	3	11·74	— 0·060	— 0·02	3396	2055	J 198	..
327	71 26 6·72	56·73	3	11·76	+ 0·401	+ 0·06	..	..	..	645
328	155 40 11·07	60·24	2	11·81	+ 0·075	+ 0·12	3384	2057	162.J 199	..
329	69 5 10·01	58·81	27	11·83	+ c·405	+ 0·06	..	..	..	646
330	167 1 52·38	58·84	6	11·84	— 0·195	— 0·01	3435	2073	J 200	..
331	69 30 2·50	60·96	2	12·35	+ 0·393	0·00	..	..	..	..
332	69 32 14·46	58·53	3	12·36	+ 0·393	— 0·04	..	..	..	655
333	132 30 2·91	60·22	9	12·40	+ 0·237	+ 0·02	3446	2114	..	..
334	101 58 58·64	57·30	1	12·44	+ 0·320	+ 0·03	..	..	..	1070*
335	124 48 48·20	60·20	2	12·53	+ 0·262	+ 0·10	3462	2127	163.J 201	657
336	68 1 50·67	57·15	8	12·57	+ 0·392	— 0·01	..	..	..	659
337	136 9 8·02	60·28	1	12·62	+ 0·220	+ 0·02	3470	2141	J 202	..
338	142 25 33·61	60·18	2	12·64	+ 0·190	— 0·02	3482	2148	J 203	..
339	71 20 1·64	59·14	13	12·67	+ 0·382	+ 0·24	..	..	..	666
340	149 15 45·20	60·12	1	12·73	+ 0·145	+ 0·02	3504	2163	J 205	..
341	83 4 12·28	58·71	19	12·85	+ 0·353	+ 0·04	..	..	..	671
342	91 23 11·69	57·22	2	12·90	+ 0·335	..	..	..	..	..
343	144 11 48·80	60·21	1	12·95	+ 0·179	+ 0·09	3532	2194	167.J 206	H 44
344	135 31 50·50	60·18	3	12·97	+ 0·220	— 0·04	3526	2198	J 207	..
345	92 55 34·30	58·23	3	13·05	+ 0·329	0·00	..	..	..	1090*
346	-71 28 43·16	57·30	3	13·07	+ 0·372	..	..	..	..	..
347	146 15 23·15	60·23	6	13·10	+ 0·166	+ 0·02	3554	2217	..	..
348	96 39 18·02	60·24	3	13·20	+ 0·318	0·00	..	..	..	1097*
349	168 27 12·86	60·21	2	13·29	— 0·206	— 0·02	3623	2254	J 208	..
350	77 50 28·72	58·08	2	+ 13·44	+ 0·350	+ 0·02	..	..	..	678

No.	No. in B.A.C.	Magnitude	Star's Name.	Mean R.A. 1860, Jan. 1.	Mean Year and Fraction of Year.	No. of Obs. of R.A.	Annual Precess. in R.A. for 1860,	Secular Variation of Precess. in R.A.	Annual Proper Motion in R.A.
				b m s	1800		s	s	s
351	3055	4	65 Cancri............ a	8 50 50	..	..	+ 3·288	— 0·0098	0·000
352	3058	7	Piazzi VIII. 224..	8 51 16·10	57·30	1	3·404	— 0·0131	..
353	3065	6	Piazzi VIII. 227..	8 52 10	..	..	2·802	+ 0·0002	..
354	3073	4	Carinæ............ b¹	8 53 32·70	60·22	11	1·475	— 0·0052	— 0·002
355	3074	7	68 Cancri.........	8 53 51·99	57·31	1	3·380	— 0·0126	0·000
356	3079	6	69 Cancri........ ν	8 54 32·80	58·08	2	3·523	— 0·0172	— 0·003
357	3089	4	Carinæ............ b²	8 55 58·03	60·28	4	1·499	— 0·0047	— 0·020
358	3110	5	Velorum............ c	8 59 19·47	60·16	1	2·071	+ 0·0037	— 0·018
359	3111	5	76 Cancri........ ε	9 0 9·69	60·72	3	3·260	— 0·0094	— 0·002
360	3115	7	78 Cancri	9 1 11·11	57·31	1	3·378	— 0·0130	— 0·001
361	3117	5	77 Cancri .... ξ	9 1 18·25	57·03	8	3·464	— 0·0159	— 0·002
362	3123	6	79 Cancri..........	9 2 17·93	59·55	3	3·461	— 0·0159	+ 0·002
363	3126	3	Argûs............... λ	9 2 50·82	60·26	3	2·205	+ 0·0045	— 0·006
364	3132	6·7	81 Cancri........ π¹	9 4 37·76	59·13	1	3·330	— 0·0117	— 0·033
365	3136	5	Carinæ............ G	9 4 44·94	60·26	1	0·216	— 0·0605	— 0·033
366	3138	6	Bradley 1299	9 5 37·20	58·42	3	3·442	— 0·0155	+ 0·005
367	3147	6	82 Cancri .... π²	9 7 29·94	58·47	3	3·326	— 0·0118	— 0·001
368	3152	5	Carinæ............ i	9 8 5·75	60·25	1	1·376	— 0·0081	— 0·019
369	3161	6	24 Hydræ...........	9 9 (49·86)	..	..	2·942	— 0·0016	— 0·003
370	3163	5	Velorum............ l	9 10 5·60	60·21	1	2·367	+ 0·0051	— 0·015
371	..	10	....	9 10 52·83	56·03	4	3·229	— 0·0089	..
372	3171	6	83 Cancri...........	9 11 9·70	58·97	7	3·369	— 0·0134	— 0·012
373	3177	1	Argûs.............. β	9 11 38·83	58·90	12	0·720	— 0·0345	— 0·032
374	3186	2	Argûs............... ι	9 13 20·49	60·23	3	1·611	— 0·0022	— 0·003
375	3187	5	Velorum.......... K	9 13 26·46	60·28	1	1·996	+ 0·0041	— 0·006
376	3195	6·5	Mali............... h	9 15 17·82	60·26	1	+ 2·655	+ 0·0034	+ 0·010
377	3211	5·6	Octantis............ ζ	9 16 16·33	56·33	10	— 7·139	— 1·4857	— 0·069
378	3213	3	Argûs.............. κ	9 17 46·69	60·22	3	+ 1·857	+ 0·0027	— 0·007
379	3223	2	30 Hydræ......... a	9 20 42·44	59·59	17	2·951	— 0·0015	— 0·004
380	3246	5·4	4 Leonis........... λ	9 23 43·59	57·45	11	3·440	— 0·0172	— 0·004
381	3257	4	Argûs............ ψ	9 25 11·18	60·16	2	2·375	+ 0·0064	— 0·027
382	3269	5	Velorum.......... N	9 26 57·96	60·21	2	+ 1·825	+ 0·0028	— 0·015
383	3279	5·6	Chamæleontis.......	9 28 40·47	56·26	5	— 1·666	— 0·2845	..
384	..	11	....	9 32 2·52	56·03	5	+ 3·168	— 0·0073	..
385	3312	4·3	14 Leonis .... o	9 33 40·56	60·10	1	+ 3·220	— 0·0093	— 0·013

No.	Mean N.P.D. 1860, Jan. 1.	Mean Year and Fraction of Year.	No. of Obs. of N.P.D.	Annual Precess. in N.P.D.	Secular Variation of Precess. in N.P.D.	Annual Proper Motion in N.P.D.	No. for reference.			
							Lacaille.	Brisbane.	Fallows or Johnson.	Greenwich or Henderson.
	° ′ ″	1800		″	″	″				
351	77 36 10·67	60·92	2	+ 13·60	+ 0·347	+ 0·04	··	2268	168	683
352	71 19 21·37	57·30	1	13·63	+ 0·359	··	··	··	··	··
353	105 36 3·30	60·12	1	13·69	+ 0·293	··	··	··	··	··
354	148 41 23·34	60·22	11	13·78	+ 0·150	− 0·04	3639	2293	169.J 210	··
355	72 22 20·37	57·31	1	13·80	+ 0·352	0·00	··	··	··	1119*
356	64 59 55·36	58·08	2	13·84	+ 0·366	+ 0·01	··	··	··	687
357	148 32 56·10	60·29	7	13·93	+ 0·151	− 0·24	3661	2311	170.J 211	··
358	136 32 26·92	60·16	1	14·14	+ 0·208	− 0·14	3677	2326	J 212	··
359	78 46 14·62	60·72	3	14·19	+ 0·330	0,00	··	··	171	696
360	71 57 55·55	57·31	1	14·25	+ 0·340	− 0·01	··	··	··	1130*
361	67 23 26·61	57·95	9	14·26	+ 0·349	− 0·01	··	··	··	698
362	67 26 14·76	59·55	3	14·32	+ 0·347	+ 0·03	··	··	··	699
363	132 52 8·11	60·26	3	14·36	+ 0·218	0·00	3699	2346	172.J 214	H 63
364	74 26 30·98	59·13	1	14·47	+ 0·330	− 0·28	··	2356	··	702
365	162 2 21·42	60·26	1	14·47	+ 0·015	− 0·02	3736	2374	J 215	··
366	68 8 32·88	58·42	3	14·53	+ 0·340	+ 0·01	··	··	··	704
367	74 28 48·11	58·47	3	14·64	+ 0·325	− 0·02	··	2384	··	708
368	151 44 35·95	60·25	1	14·67	+ 0·130	0·00	3753	2394	J 217	··
369	98 9 45·33	60·18	1	14·78	+ 0·284	− 0·05	··	··	··	1139*
370	127 59 18·48	60·21	1	14·79	+ 0·227	+ 0·08	3756	2407	··	··
371	80 6 54·82	56·03	4	14·84	+ 0·310	··	··	··	··	··
372	71 42 11·64	58·68	31	14·86	+ 0·324	+ 0·16	··	··	··	711
373	159 8 27·22	58·72	13	14·88	+ 0·064	− 0·09	3791	2425	174.J 218	H 16
374	148 41 20·55	60·23	3	14·98	+ 0·150	+ 0·02	3792	2429	175.J 219	H 35
375	140 27 49·08	60·29	1	14·99	+ 0·187	− 0·02	3786	2428	··	··
376	115 22 19·05	60·26	1	15·10	+ 0·248	− 0·11	3793	2436	176.J 220	713
377	175 5 48·85	56·33	10	15·15	− 0·689	− 0·01	3953	2491	··	··
378	144 24 49·96	60·23	6	15·24	+ 0·170	− 0·01	3816	2459	177.J 221	H 43
379	98 3 13·30	58·47	67	15·40	+ 0·269	− 0·03	··	2478	178.J 222	722
380	66 25 0·28	57·36	12	15·57	+ 0·310	+ 0·04	··	··	··	729
381	129 51 17·76	60·16	2	15·65	+ 0·210	− 0·07	3885	2519	J 224	··
382	146 25 4·61	60·21	2	15·75	+ 0·158	+ 0·01	3910	2535	J 225	··
383	170 10 50·66	56·26	5	15·84	− 0·150	− 0·11	3981	2568	··	··
384	83 10 46·92	56·03	5	16·02	+ 0·271	··	··	··	··	··
385	79 28 21·41	59·77	2	16·11	+ 0·273	+ 0·04	··	2586	180	747

No.	No. in B.A.C.	Magnitude	Star's Name.	Mean R.A. 1860, Jan. 1.	Mean Year and Fraction of Year.	No. of Obs. of R.A.	Annual Precess. in R.A. for 1860.	Secular Variation of Precess. in R.A.	Annual Proper Motion in R.A.
				h m s	1800				
386	3320	5	Carinæ .... *m*	9 35 28·24	60·25	3	+ 1·667	0·0000	..
387	3321	6	16 Leonis........ ψ	9 36 6·21	57·03	1	+ 3·278	— 0·0115	— 0·002
388	3334	5·6	Chamæleontis..... ζ	9 37 52·01	56·27	4	— 1·479	— 0·2817	+ 0·001
389	3331	3	17 Leonis......... ε	9 37 53·90	59·29	3	+ 3·425	— 0·0180	— 0·004
390	..	7	Lacaille 3999........	9 39 22·58	60·29	4	2·590	+ 0·0065	..
391	3345	Var.	Bradley 1373........	9 40 1·51	59·20	1	3·236	— 0·0101	+ 0·002
392	..	8	W.B. IX. 888	9 40 51·47	56·02	4	3·147	— 0·0069	..
393	3353	5	Carinæ............... *l*	9 41 23·96	60·28	4	1·651	— 0·0001	— 0·011
394	3365	3	Argûs............... υ	9 43 36·08	60·27	9	1·506	— 0·0045	0·000
395	..	11	....	9 48 20·24	56·03	4	3·129	— 0·0062	..
396	3406	·5	27 Leonis .... ν	9 50 41·29	58·64	2	3·239	— 0·0106	— 0·004
397	3410	4	Argûs.............. φ	9 51 57·04	60·27	9	2·100	+ 0·0093	— 0·007
398	3415	5	29 Leonis......... π	9 52 48·75	59·81	13	3·180	— 0·0081	— 0·003
399	..	9·10	....	9 55 28·90	56·02	4	3·110	— 0·0054	..
400	3453	3·4	30 Leonis......... η	9 59 41·71	57·15	4	3·283	— 0·0131	— 0·004
401	3459	1·2	32 Leonis......... α	10 0 54·77	59·33	32	3·221	— 0·0102	— 0·019
402	..	6·7	....	10 2 20·55	60·23	2	+ 2·687	+ 0·0080	..
403	3480	5·6	Chamæleontis.... μ¹	10 4 18·98	56·33	5	— 1·273	— 0·3230	..
404	3493	5·6	Chamæleontis.... μ²	10 6 40·89	56·37	2	— 0·882	· 0·2599	..
405	3492	·6	21 Sextantis.........	10 7 10	..	..	+ 2·991	— 0·0003	+ 0·002
406	3509	4	Velorum .... *q*	10 8 51·85	60·30	7	2·522	+ 0·0118	— 0·017
407	3516	4	Argûs......,...... ω	10 10 23·99	57·82	8	1·440	— 0·0070	— 0·026
408	3523	2	41 Leonis......... γ¹	10 12 14·97	58·90	5	3·299	— 0·0149	+ 0·019
409	3526	5	Carinæ............ *g*	10 12 24·60	60·26	4	1·997	+ 0·0114	— 0·014
410	3536	5	Velorum .... *V*	10 14 21·33	60·28	3	2·244	+ 0·0142	— 0·013
411	3546	5	Velorum.......... T	10 15 42·44	60·31	2	2·222	+ 0·0144	— 0·013
412	3550	6·7	24 Sextantis.........	10 16 18·35	56·03	5	3·070	— 0·0031	+ 0·001
413	3552	5	Velorum........... *r*	10 16 19·61	60·19	1	2·565	+ 0·0126	— 0·006
414	3561	6	44 Leonis...........	10 17 52·31	58·08	4	3·168	— 0·0080	— 0·007
415	3568	4	42 Hydræ..... μ	10 19 19	..	..	2·908	+ 0·0039	— 0·010
416	3575	6	45 Leonis...........	10 20 15·12	58·76	10	3·176	— 0·0085	— 0·002
417	3578	4	Antliæ............. α	10 20 44·88	60·28	3	2·743	+ 0·0096	— 0·005
418	3579	6	Piazzi X. 83.........	10 21 19·24	57·03	2	3·222	— 0·0111	— 0·001
419	3585	4·5	Carinæ............. *I*	10 21 36·54	60·32	1	1·215	— 0·0213	— 0·008
420	3589	5	Velorum.......... P	10 22 11·95	60·31	1	+ 2·223	+ 0·0160	..

No.	Mean N.P.D. 1860, Jan. 1.	Mean Year and Fraction of Year.	No. of Obs. of N.P.D.	Annual Precess. in N.P.D. for 1860.	Secular Variation of Precess. in N.P.D.	Annual Proper Motion in N.P.D.	No. for reference.			
							Lacaille.	Brisbane.	Fallows or Johnson.	Greenwich or Henderson.
	° ′ ″	1800		″	″	″				
386	150 41 43·23	60·25	3	+ 16·20	+ 0·136	0·00	3987	2607	..	..
387	75 20 22·72	57·03	1	16·23	+ 0·274	+ 0·02	..	..	..	750
388	170 18 40·19	56·26	5	16·32	− 0·132	+ 0·03	4048	2648	..	..
389	65 34 58·92	59·49	12	16·32	+ 0·283	+ 0·02	..	2620	..	751
390	122 2 18·60	60·28	3	16·40	+ 0·211	..	..	..	..	..
391	77 55 24·49	59·20	1	16·43	+ 0·264	+ 0·17	..	..	..	756
392	84 23 39·74	56·02	4	16·47	+ 0·255	..	..	..	..	..
393	151 51 46·82	60·28	4	16·50	+ 0·130	− 0·03	4033	2664	J 229	..
394	154 25 24·33	60·27	8	16·61	+ 0·116	+ 0·01	4051	2682	181.J 230	H 21
395	85 30 28·79	56·03	4	16·84	+ 0·241	..	..	..	..	..
396	76 53 20·05	58·77	20	16·95	+ 0·245	+ 0·01	..	..	..	767
397	143 54 8·67	60·27	10	17·01	+ 0·155	+ 0·01	4093	2752	J 232	..
398	81 17 8·43	59·75	27	17·05	+ 0·237	+ 0·03	..	2757	182	768
399	86 48 44·85	56·02	4	17·17	+ 0·227	..	..	..	..	..
400	72 33 22·07	56·96	7	17·36	+ 0·233	0·00	..	..	..	771
401	77 21 0·29	58·96	88	17·41	+ 0·226	− 0·01	..	2838	184	775
402	120 25 5·85	60·23	2	17·47	+ 0·184	..	..	..	..	..
403	171 32 10·33	56·33	5	17·55	− 0·098	0·00	4232	2880	..	..
404	170 52 59·06	56·37	2	17·65	− 0·069	− 0·03	4246	2901	..	..
405	97 17 59·22	60·18	1	17·68	+ 0·198	+ 0·02	..	..	..	1241*
406	131 25 45·47	60·25	7	17·74	+ 0·163	− 0·05	4212	2904	J 234	..
407	159 20 36·41	57·82	8	17·81	+ 0·089	+ 0·01	4243	2924	J 235	..
408	69 27 6·11	58·70	20	17·88	+ 0·210	+ 0·15	..	2929	..	791
409	150 38 0·77	60·26	4	17·89	+ 0·124	− 0·02	4249	2935	J 236	..
410	144 19 37·84	60·28	3	17·96	+ 0·138	+ 0·04	4263	2952	..	..
411	145 20 21·72	60·31	2	18·01	+ 0·134	+ 0·05	4272	2972	J 237	..
412	90 11 39·49	56·03	5	18·04	+ 0·189	+ 0·03	..	..	..	..
413	130 56 48·11	60·19	1	18·04	+ 0·156	− 0·03	4271	2974	J 238	..
414	80 30 16·90	58·08	4	18·10	+ 0·191	+ 0·12	..	2984	188	802
415	106 7 22·09	60·26	1	18·15	+ 0·173	+ 0·11	..	..	J 239	804
416	79 31 30·62	58·76	10	18·19	+ 0·188	+ 0·01	..	..	..	808
417	120 21 21·73	60·28	3	18·21	+ 0·160	+ 0·05	4298	3011	J 240	809
418	74 56 33·12	57·03	2	18·23	+ 0·188	+ 0·04	..	..	..	811
419	163 19 10·56	60·32	1	18·24	+ 0·066	0·00	4319	3025	J 241	..
420	146 55 31·67	60·30	3	+ 18·26	+ 0·126	− 0·03	4310	3023	..	..

No.	No. in B.A.C.	Magnitude	Star's Name.	Mean R.A. 1860, Jan. 1.	Mean year and Fraction of year.	No. of Obs. R.A.	Annual Precess. in R.A. for 1860.	Secular Variation of Precess. in R.A.	Annual Proper Motion in R.A.
				h m s	1800				
421	3594	5	Carinæ........ ..... s	10 22 44·82	60·29	3	+2·190	+0·0160	+0·003
422	3609	4	47 Leonis......... p	10 25 26·27	58·60	12	3·167	−0·0081	0·000
423	3619	4	Carinæ............ p	10 27 3·10	60·24	7	2·122	+0·0166	−0·012
424	3644	5	Velorum.......... p	10 31 25·37	60·29	6	2·522	+0·0170	−0·017
425	..	11	....	10 32 42·15	56·03	5	3·048	−0·0013	..
426	3655	5	Carinæ............ f²	10 33 25·88	60·30	2	2·269	+0·0193	0·000
427	3660	5	Chamæleontis.... γ	10 33 47·36	60·31	1	0·784	−0·0649	−0·012
428	3681	5·6	Brisbane 3176......	10 37 16·62	60·23	1	2·115	+0·0194	−0·002
429	3686	3	Argûs............ θ	10 37 58·08	60·31	5	2·126	+0·0197	−0·011
430	3690	6	37 Sextantis.........	10 38 48·20	60·99	1	3·130	−0·0060	−0·002
431	3695	Var.	Argûs............ η	10 39 38·20	59·84	9	2·309	+0·0214	−0·003
432	..	10	....	10 39 43·23	56·02	4	3·040	−0·0004	..
433	3702	3	Argûs............ μ	10 40 45·38	60·28	5	2·557	+0·0192	+0·002
434	3708	5	53 Leonis......... l	10 41 53·77	58·89	6	3·161	−0·0082	−0·003
435	3723	5·6	Chamæleontis.... δ¹	10 43 53·93	56·32	5	0·667	−0·0894	..
436	3724	5	Chamæleontis.... δ²	10 44 25·40	57·03	6	0·666	−0·0895	−0·040
437	3740	5	Carinæ............ u	10 47 48·95	60·29	7	2·406	+0·0243	0·000
438	3755	5·6	Lacaille 4527......	10 50 11·95	60·26	5	2·777	+0·0153	−0·005
439	3768	5	58 Leonis......... d	10 53 19·78	60·25	2	3·101	−0·0039	−0·002
440	3769	5	59 Leonis.......... c	10 53 29·34	56·36	1	3·118	−0·0052	−0·005
441	..	10	....	10 55 59·63	56·03	4	3·030	+0·0016	..
442	3788	5	63 Leonis......... χ	10 57 47·58	58·64	16	3·123	−0·0057	−0·024
443	3793	4·5	9 Crateris.........	10 58 35·44	60·29	6	2·895	+0·0114	−0·009
444	3794	5	Bradley 1538.......	10 59 10·50	60·30	5	+2·897	+0·0114	+0·005
445	3803	6	Octantis.......... η	11 0 10·69	56·35	5	−0·126	−0·2893	..
446	..	10	....	11 4 0·69	56·03	4	+3·030	+0·0025	..
447	3832	5	69 Leonis......... p⁰	11 6 35·63	60·26	4	3·076	−0·0014	0·000
448	3834	2·3	68 Leonis......... δ	11 6 39·48	60·01	4	3·192	−0·0133	+0·011
449	3836	6	B.F. 1589..........	11 6 41·85	58·38	2	3·088	−0·0026	..
450	3843	6	73 Leonis......... s	11 8 32·22	56·21	2	3·147	−0·0086	−0·001
451	3848	5·4	74 Leonis .... φ	11 9 32·69	59·97	5	3·057	+0·0006	−0·009
452	3859	3·4	12 Hydræ etCrateris δ	11 12 20·60	59·19	15	3·003	+0·0063	−0·009
453	3862	4	77 Leonis .... σ	11 13 54·98	58·43	7	3·104	−0·0042	−0·009
454	3866	4	Centauri.......... π	11 14 38·02	60·27	6	2·714	+0·0301	−0·005
455	3877	4	78 Leonis.......... ι	11 16 37·35	56·29	3	+3·122	−0·0066	+0·007

No.	Mean N.P.D. 1860. Jan. 1.	Mean Year and Fraction of Year.	No. of Obs. of N.P.D.	Annual Precess. in N.P.D. for 1860.	Secular Variation of Precess. in N.P.D	Annual Proper Motion in N.P.D.	No. for reference.			
	o ′ ″			′	″	″	Lacaille.	Brisbane.	Fallows or Johnson.	Greenwich or Henderson.
421	148 1 31·55	1800 60·29	3	+ 18·28	+ 0·124	+ 0·03	4314	3031	··	··
422	79 58 26·93	58·65	33	18·37	+ 0·177	+ 0·03	···	3046	··	819
423	150 57 57·66	60·24	8	18·43	+ 0·114	+ 0·02	4348	3072	J 242	··
424	137 29 56·98	60·29	7	18·58	+ 0·131	+ 0·03	4378	3114	J 243	··
425	92 47 37·64	56·03	5	18·62	+ 0·158	··	··	··	··	··
426	148 27 16·67	60·30	2	18·64	+ 0·114	− 0·10	4396	3127	··	··
427	167 52 56·13	60·31	1	18·65	+ 0·034	0·00	4428	3146	J 245	··
428	153 44 5·23	60·23	1	18·76	+ 0·101	+ 0·05	··	3176	J 246	··
429	153 39 41·94	60·31	5	18·78	+ 0·101	+ 0·02	4447	3184	192.J 247	H 22
430	82 53 23·89	60·99	1	18·81	+ 0·150	+ 0·06	··	··	193	836
431	148 56 57·20	59·57	11	18·84	+ 0·108	+ 0·01	4457	3198	194.J 248	H 33
432	93 58 40·06	56·02	4	18·84	+ 0·145	··	··	··	··	··
433	138 40 53·43	60·29	5	18·87	+ 0·118	+ 0·08	4461	3206	195.J 249	H 50
434	78 42 53·86	58·98	10	18·90	+ 0·146	+ 0·02	··	··	··	840
435	169 43 50·28	56·32	5	18·96	+ 0·023	+ 0·08	4509	3243	··	··
436	169 48 7·08	57·03	6	18·98	+ 0·024	+ 0·01	4513	3247	J 251	··
437	148 6 37·06	60·30	8	19·07	+ 0·101	+ 0·02	4515	3274	J 252	··
438	126 23 8·93	60·26	5	19·13	+ 0·113	+ 0·20	4527	3293	··	848
439	85 37 53·38	60·25	2	19·21	+ 0·122	+ 0·03	··	··	··	851
440	83 8 49·69	58·20	9	19·22	+ 0·122	+ 0·06	··	··	197	852
441	96 28 8·37	56·03	4	19·28	+ 0·114	··	··	··	··	··
442	81 54 28·56	58·78	43	19·32	+ 0·114	+ 0·08	··	··	199	860
443	116 32 19·57	60·29	6	19·34	+ 0·104	+ 0·03	4583	3376	J 254	861
444	116 31 54·86	60·30	5	19·35	+ 0·103	+ 0·03	4587	3382	··	863
445	173 50 26·80	56·35	5	19·38	− 0·013	+ 0·08	4643	3409	··	··
446	97 29 3·05	56·03	4	19·46	+ 0·099	··	··	··	··	··
447	89 18 30·19	60·28	3	19·51	+ 0·095	0·00	··	3456	··	872
448	68 42 35·40	58·72	9	19·52	+ 0·099	+ 0·14	··	··	··	873
449	86 58 6·92	58·38	2	19·52	+ 0·095	··	··	··	··	874
450	75 55 44·98	56·21	2	19·55	+ 0·094	+ 0·04	··	··	··	878
451	92 53 11·97	59·82	5	19·57	+ 0·089	+ 0·04	··	··	J 257	879
452	104 1 16·82	58·53	71	19·62	+ 0·082	− 0·18	··	··	J 258	884
453	83 12 14·05	58·39	19	19·65	+ 0·082	+ 0·03	··	··	203	885
454	143 43 27·41	60·28	8	19·66	+ 0·069	+ 0·03	4717	3544	204.J 259	··
455	78 41 59·85	56·33	4	+ 19·70	+ 0·077	+ 0·07	··	··	··	888

No.	No. in B.A.C.	Magnitude	Star's Name.	Mean R.A. 1860, Jan. 1.	Mean Year and Fraction of Year.	No. of Obs. of R.A.	Annual Precess. in R.A. for 1860.	Secular Variation of Precess. in R.A.	Annual Proper Motion in R.A.
				h m s	1800				
456	..	9'10	....	11 19 13'79	56'02	4	+3'030	+0'0047	..
457	3900	5	84 Leonis ,......... r	11 20 44'21	58'07	6	3'086	—0'0022	—0'001
458	3916	5	87 Leonis......... e	11 23 9'70	59'63	7	3'064	+0'0010	—0'001
459	3921	6'7	17 Crateris, 1st Star	11 25 20'06	60'29	2	2'963	+0'0149	—0'005
460	3922	5	17 Crateris, 2nd Star	11 25 20'35	60'25	4	2'963	+0'0149	—0'005
461	3928	4	Lacaille 4779.......	11 26 7'30	60'30	4	2'953	+0'0165	—0'016
462	3930	6	89 Leonis............	11 27 12'08	57'66	3	3'085	—0'0019	—0'008
463	..	7	W.B. XI. 475	11 27 50'13	56'02	4	3'035	+0'0058	..
464	3941	4	Centauri.......... λ	11 29 20'46	60'34	2	2'733	+0'0442	—0'010
465	3946	5'4	91 Leonis......... υ	11 29 46'87	58'11	5	3'072	+0'0002	—0'003
466	..	8	Lalande 22032......	11 30 6'71	60'21	2	3'073	+0'0001	—0'004
467	..	9	Lalande 22038......	11 30 20'21	60'29	10	3'073	+0'0001	—0'004
468	..	10'11	....	11 36 56'04	56'03	5	3'043	+0'0069	..
469	3982	4'5	3 Virginis......... ν	11 38 39'79	56'20	4	3'088	—0'0032	+0'001
470	3995	2	94 Leonis......... β	11 41 54'92	60'31	4	3'101	—0'0075	—0'036
471	4002	3'4	5 Virginis......... β	11 43 24'12	56'89	5	3'076	—0'0004	+0'048
472	4006	6	Piazzi XI. 167.....	11 43 52'86	59'96	4	3'065	+0'0034	+0'011
473	4015	4	Lacaille 4923........	11 45 50'64	60'29	7	3'018	+0'0199	—0'008
474	4017	2'3	64 Ursæ Majoris γ	11 46 27	..	..	3'183	—0'0437	+0'011
475	4048	5	Chamæleontis..... e	11 52 43'73	60'30	5	2'882	+0'1190	—0'018
476	4049	6	7 Virginis .... b	11 52 46'60	56'89	4	3'075	—0'0008	—0'002
477	4052	4'5	8 Virginis......... π	11 53 41'94	56'44	1	3'077	—0'0023	0'000
478	4061	5'6	Crucis............. θ¹	11 55 55'21	60'28	9	3'027	+0'0574	—0'025
479	4067	5 6	Crucis............. θ²	11 57 8'19	60'35	1	3'040	+0'0577	+0'002
480	4085	6	Lacaille 5029........	12 0 50'96	60'31	1	3'078	+0'0373	—0'011
481	4087	3	Centauri............ δ	12 1 7'17	40'28	5	3'080	+0'0373	..
482	4090	4	1 Corvi............ α	12 1 11'74	60'28	2	3'075	+0'0153	+0'010
483	4094	6	10 Virginis..........	12 2 30'90	57'56	9	3'071	+0'0006	+0'001
484	4097	3	2 Corvi............ e	12 2 55'78	59'39	13	3'079	+0'0141	—0'005
485	4120	3	Crucis............. δ	12 7 43'88	60'34	2	3'144	+0'0522	—0'010
486	4131	5	Chamæleontis.... β	12 10 12'99	56'39	12	3'366	+0'1778	—0'017
487	4137	6	13 Virginis..........	12 11 29'64	60'26	2	3'072	+0'0026	0'000
488	4145	3'4	15 Virginis........ η	12 12 44'70	57'88	5	3'072	+0'0026	—0'007
489	4187	1	Crucis............. α	12 18 50'79	60'26	1	3'281	+0'0673	—0'022
490	4189	6	Centauri............	12 18 59'16	57'46	1	+3'207	+0'0442	—0'002

481. The R.A. has been brought up with Precession alone.

No.	Mean N.P.D. 1860, Jan. 1.	Mean Year and Fraction of Year.	No. of Obs. of N.P.D.	Annual Precess. in N.P.D. for 1860.	Secular Variation of Precess.in N.P.D.	Annual Proper Motion in N.P.D.	No. for reference.			
							Lacaille.	Brisbane.	Fallows or Johnson.	Greenwich or Henderson.
	° ′ ″	1800		″	″	″				
456	99 58 39·97	56·02	4	+ 19·74	+ 0·070	··	··	··	··	··
457	86 22 23·21	57·79	16	19·76	+ 0·068	+ 0·02	··	··	··	897
458	92 13 53·17	59·21	7	19·80	+ 0·063	+ 0·03	··	··	J 262	903
459	118 29 46·94	60·29	2	19·83	+ 0·057	− 0·11	··	··	··	905
460	118 29 39·24	60·25	4	19·83	+ 0·057	− 0·11	4770	3628	··	906
461	121 4 59·54	60·32	3	19·84	+ 0·055	+ 0·03	4779	3641	J 263	907
462	86 9 45·47	57·66	3	19·85	+ 0·056	+ 0·13	··	··	208	908
463	101 18 53·94	56·02	4	19·86	+ 0·053	··	··	··	··	··
464	152 14 43·97	60·34	2	19·88	+ 0·044	+ 0·06	4804	3669	209.J 264	··
465	90 3 3·53	58·48	22	19·88	+ 0·050	− 0·03	··	··	210	915
466	89 48 13·66	60·21	2	19·89	+ 0·050	+ 0·02	··	··	··	··
467	89 46 17·98	60·29	10	19·89	+ 0·049	+ 0·12	··	··	··	··
468	102 21 19·20	56·03	5	19·95	+ 0·036	··	··	··	··	··
469	82 41 10·37	56·17	5	19·97	+ 0·033	+ 0·21	··	··	··	927
470	74 38 43·56	58·23	14	19·99	+ 0·027	+ 0·10	··	3780	211	931
471	87 26 47·04	57·01	19	20·00	+ 0·024	+ 0·28	··	3791	212	932
472	94 33 17·23	59·96	4	20·01	+·0·023	+ 0·10	··	··	··	933
473	123 7 45·34	60·29	7	20·02	+ 0·019	+ 0·05	4923	3811	213.J 267	935
474	35 31 13·73	56·48	1	20·02	+ 0·019	0·00	··	··	··	937
475	167 26 31·72	60·30	5	20·05	+ 0·004	+ 0·02	4974	3865	J 268	··
476	85 33 54·26	56·89	4	20·05	+ 0·006	+ 0·02	··	··	··	945
477	82 36 17·31	56·44	2	20·05	+ 0·004	+ 0·04	··	··	··	947
478	152 32 0·19	60·28	9	20·05	− 0·001	+ 0·06	4990	3892	··	··
479	152 23 10·30	60·35	1	20·05	− 0·003	+ 0·05	4999	3901	··	··
480	139 52 52·85	60·31	1	20·06	− 0·011	+ 0·02	5029	3930	216	··
481	139 56 33·98	60·32	1	20·06	− 0·011	+ 0·07	5033	3934	217.J 270	H 48
482	113 56 53·41	60·28	2	20·06	− 0·011	+ 0·04	5035	··	J 271	955
483	87 18 56·82	57·56	9	20·05	− 0·014	+ 0·21	··	··	··	956
484	111 50 27·76	59·16	19	20·05	− 0·014	− 0·01	··	··	J 272	957
485	147 58 11·81	60·34	2	20·04	− 0·025	+ 0·06	5075	3975	218.J 274	H 37
486	168 32 4·68	56·40	10	20·04	− 0·031	− 0·02	5085	3986	220.J 276	H 13
487	90 0 31·60	60·26	2	20·03	− 0·031	+ 0·04	··	··	221	969
488	89 53 18·32	57·94	22	20·03	− 0·034	+ 0·03	··	··	··	972
489	152 19 24·27	60·26	1	19·99	− 0·048	− 0·05	5148	4050	223.J 279	H 26
490	140 40 28·23	57·46	1	+ 19·99	− 0·047	+ 0·10	5150	4052	··	··

No.	No. in B.A.C.	Magnitude	Star's Name.	Mean R.A. 1860, Jan. 1.	Mean Year and Fraction of Year.	No. of Obs. of R.A.	Annual Precess. in R.A. for 1860.	Secular Variation of Precess. in R.A.	Annual Proper Motion in R.A.
				h m s	1800				
491	..	7·8	Lalande 23305....,	12 20 40·48	56·09	5	+3·111	+0·0128	+0·003
492	4230	6	21 Virginis........ g	12 26 33·35	59·05	6	3·096	+0·0080	−0·009
493	4234	2·3	9 Corvi......... β	12 27 2·35	59·12	35	3·138	+0·0163	−0·008
494	4237	7	Virg, Piazzi XII.125	12 27 13	..	..	3·074	+0·0038	−0·009
495	..	10	....	12 28 37·11	56·12	4	3·129	+0·0139	−0·020
496	4247	6	25 Virginis...... f	12 29 34·84	58·05	6	3·087	+0·0062	0·000
497	..	7	Brisbane 4091.......	12 29 50·92	58·48	57	13·329	+13·4285	−0·020
498	4257	5	26 Virginis........ χ	12 32 1·39	59·33	2	3·096	+0·0075	0·000
499	4268	3·2	29 Virginis....... γ^1	12 34 34·00	59·23	11	3·074	+0·0042	−0·037
500	..	3·2	29 Virginis....... γ^2	..	..	..	3·074	+0·0042	−0·037
501	4269	6	28 Virginis, ...	12 34 43·47	58·63	6	3·096	+0·0074	+0·005
502	4275	6	Crucis...............	12 35 12·72	57·46	1	3·369	+0·0545	−0·004
503	..	11·12	....	12 35 15·03	56·11	5	3·146	+0·0148	..
504	4277	6	W.B. XII. 603.....	12 36 (26·70)	..	..	3·075	+0·0043	..
505	4280	4	Muscæ........... β	12 37 44·19	60·33	1	3·597	+0·0986	−0·014
506	4289	2	Crucis.... .... β	12 39 34·10	60·40	3	3·453	+0·0649	−0·009
507	4293	5	Octantis............ ι	12 40 40·49	56·41	5	5·463	+0·7644	..
508	4313	5·6	Centauri.............	12 44 15·12	57·46	1	3·279	+0·0310	−0·007
509	..	12	....	12 44 45·10	56·13	4	3·170	+0·0160	..
510	4321	5	Lacaille 5312.......	12 45 41·66	60·35	5	3·290	+0·0318	−0·002
511	4323	6	38 Virginis.........	12 46 1·14	56·60	4	3·085	+0·0059	−0·016
512	4325	5	Centauri,.............	12 46 23·66	57·48	1	3·477	+0·0596	+0·001
513	4330	5	40 Virginis....... ψ	12 47 4·63	58·84	4	3·114	+0·0091	−0·002
514	4340	3	43 Virginis........ δ	12 48 33·09	56·44	1	3·052	+0·0025	−0·030
515	4346	3	12 Canum Venatic. a	12 49 28·28	56·43	2	2·840	−0·0154	−0·023
516	4352	6	44 Virginis......... λ	12 52 26·78	57·19	2	3·088	+0·0063	+0·002
517	4353	4	Muscæ............. δ	12 52 42·15	60·31	7	3·947	+0·1354	+0·042
518	..	10·11	....	12 53 2·65	56·12	5	3·193	+0·0171	..
519	4358	6	46 Virginis..........	12 53 23	..	..	3·086	+0·0062	+0·001
520	4368	5·6	Centauri .... ξ^1	12 55 28·39	57·48	1	3·438	+0·0455	−0·002
521	4373	6	48 Virginis,.........	12 56 41	..	..	3·081	+0·0065	−0·002
522	4379	5	Centauri.......... ξ^2	12 58 45·25	60·30	3	3·464	+0·0469	−0·016
523	4391	6	49 Virginis........ g	13 0 34·01	59·29	1	3·134	+0·0104	+0·001
524	4395	5·6	45 Hydræ........ ψ	13 1 31·24	56·11	4	3·218	+0·0182	+0·004
525	4401	4·5	51 Virginis........ θ	13 2 42·21	56·41	6	+3·102	+0·0077	−0·004

No.	Mean N.P.D. 1860. Jan. 1.	Mean Year and Fraction of Year.	No. of Obs. of N.P.D.	Annual Precess. in N.P.D. for 1860.	Secular Variation of Precess. in N.P.D.	Annual Proper Motion in N.P.D.	No. for reference.			
							Lacaille.	Brisbane.	Fallows or Johnson.	Greenwich or Henderson.
	° ′ ″	1800		″	″	″				
491	107 50 4·28	56·09	5	+19·97	−0·049	0·00	··	··	··	··
492	98 40 45·16	59·05	6	19·92	−0·061	0·00	··	··	··	986
493	112 37 19·22	58·61	98	19·92	−0·062	+0·07	··	··	227.J 285	987
494	90 38 7·97	56·32	7	19·92	−0·062	+0·04	··	··	··	989
495	108 47 26·11	56·12	4	19·90	−0·065	··	··	··	··	··
496	95 3 34·26	58·05	6	19·89	−0·067	+0·09	··	··	··	994
497	179 1 48·30	58·35	59	19·89	−0·261	−0·04	··	4091	··	··
498	97 13 27·44	59·50	7	19·86	−0·071	+0·04	··	··	··	995
499	90 40 50·66	58·52	11	19·83	−0·076	+0·05	··	4159	230.J 289	997
500	90 40 53·84	56·31	2	19·83	−0·076	+0·05	··	··	··	999
501	96 43 47·17	58·63	6	19·83	−0·077	+0·02	··	··	··	1000
502	145 24 26·96	57·46	1	19·81	−0·083	+0·07	5251	4163	··	··
503	109 45 48·29	56·11	5	19·82	−0·078	··	··	··	··	··
504	90 48 21·92	56·29	3	19·80	−0·080	··	··	··	··	1003
505	157 20 26·32	60·33	1	19·78	−0·095	+0·04	5267	4179	J 290	··
506	148 55 21·34	60·40	3	19·76	−0·095	+0·03	5277	4189	231.J 291	H 34
507	174 21 41·78	56·41	5	19·74	−0·149	+0·07	5268	4187	··	··
508	128 55 0·90	57·46	1	19·68	−0·099	−0·03	5300	4217	··	··
509	110 40 12·19	56·13	4	19·67	−0·098	··	··	··	··	··
510	129 25 0·24	60·35	5	19·66	−0·103	+0·08	5312	4232	J 292	··
511	92 47 29·58	56·41	10	19·65	−0·098	+0·03	··	··	··	1014
512	146 25 2·49	57·48	1	19·65	−0·109	+0·16	5317	4237	J 293	··
513	98 46 39·81	58·39	11	19·63	−0·101	+0·04	··	··	··	1016
514	85 50 27·96	56·44	1	19·61	−0·102	+0·09	··	··	··	1017
515	50 55 28·39	56·44	3	19·59	−0·097	−0·06	··	··	··	1019
516	93 3 20·98	56·53	7	19·53	−0·111	−0·03	··	··	233	1024
517	160 47 33·09	60·31	7	19·53	−0·139	0·00	5349	4280	J 294	··
518	111 35 41·13	56·12	5	19·52	−0·115	··	··	··	··	··
519	92 36 53·57	56·25	2	19·51	−0·112	−0·07	··	··	··	1025
520	138 46 24·93	57·48	1	19·47	−0·127	+0·06	5370	4299	··	··
521	92 54 32·15	56·23	3	19·45	−0·119	+0·02	··	··	··	1030
522	139 9 18·62	60·30	3	19·40	−0·136	+0·02	5396	4321	J 295	··
523	99 59 26·53	59·29	2	19·36	−0·128	+0·02	··	4334	··	1035
524	112 22 6·02	56·11	4	19·34	−0·133	+0·06	··	··	234.J 296	1037
525	94 47 25·73	57·15	26	+19·31	−0·131	+0·04	··	··	J 297	1039

No.	No. in B.A.C.	Magnitude	Star's Name.	Mean R.A. 1860, Jan. 1.	Mean Year and Fraction of Year.	No. of Obs. of R.A.	Annual Precess. in R.A. for 1860.	Secular Variation of Precess. in R.A.	Annual Proper Motion in R.A.
				h m s	1800		"	"	"
526	4409	5	Lacaille 5422.......	13 3 23·88	60·38	3	+ 3·408	+ 0·0373	− 0·020
527	4418	5	53 Virginis.........	13 4 36·75	60·33	1	3·175	+ 0·0138	+ 0·003
528	4426	5	Muscæ..........η	13 5 48·51	60·35	2	3·971	+ 0·1128	− 0·018
529	..	10	....	13 9 54·68	56·11	5	3·244	+ 0·0190	..
530	4442	6	58 Virginis.........	13 10 7·29	59·12	5	3·142	+ 0·0108	− 0·003
531	4460	7	Octantis..........	13 14 10·59	56·47	8	8·029	+ 1·3984	..
532	4473	7	Piazzi XIII. 67....	13 15 14·99	56·12	2	3·112	+ 0·0086	+ 0·003
533	4477	6	65 Virginis.........	13 16 (3·80)	..	..	3·104	+ 0·0080	0·000
534	4478	6	66 Virginis......-	13 17 (16·17)	..	..	3·106	+ 0·0082	+ 0·012
535	4480	1	67 Virginis........α	13 17 49·27	58·67	23	3·154	+ 0·0114	− 0·005
536	4483	5	Octantis........... κ	13 19 1·07	56·64	28	8·305	+ 1·4187	..
537	4494	5·6	69 Virginis........	13 19 59·42	59·29	1	3·196	+ ·0142	..
538	4507	4·5	Lacaille 5569.......	13 22 56·21	60·29	1	3·451	+ 0·0338	− 0·011
539	4508	7	72 Virginis....... l¹	13 23 (7·69)	..	..	3·120	+ 0·0091	+ 0·007
540	4516	5	74 Virginis....... l²	13 24 (41·38)	..	..	3·199	+ 0·0090	− 0·006
541	4517	6	Lacaille 5578........	13 24 45·45	60·41	2	3·338	+ 0·0242	− 0·007
542	4520	6	75 Virginis..........	13 25 23·12	59·53	4	3·199	+ 0·0141	− 0·001
543	4521	5	76 Virginis......... h	13 25 35·82	58·45	9	3·153	+ 0·0112	− 0·004
544	4531	6	Piazzi XIII. 126...	13 27 14·25	57·42	2	3·182	+ 0·0129	+ 0·002
545	4532	3·4	79 Virginis....... ζ	13 27 33·65	58·02	3	3·071	+ 0·0063	− 0·019
546	4546	7	81 Virginis, 1st Star	13 30 15·23	56·12	1	3·136	+ 0·0101	+ 0·003
547	4549	3	Centauri.......... ε	13 31 2·21	60·49	1	3·752	+ 0·0586	− 0·018
548	4565	6	82 Virginis........ m	13 34 16·08	56·37	1	3·147	+ 0·0107	− 0·010
549	4574	6	83 Virginis..........	13 36 57·04	58·54	1	3·224	+ 0·0150	+ 0·004
550	4579	4·5	1 Centauri......... i	13 37 44·50	60·37	3	3·422	+ 0·0278	− 0·038
551	4582	6	85 Virginis..........	13 38 3·00	58·28	4	3·221	+ 0·0148	0·000
552	4585	6	86 Virginis..........	13 38 28·82	56·15	1	3·188	+ 0·0128	− 0·004
553	..	10·11	....	13 41 10·61	56·12	5	3·350	+ 0·0222	..
554	4602	3·4	Centauri......... μ	13 41 11·96	60·47	1	3·582	+ 0·0388	+ 0·001
555	4603	5	2 Centauri......... g	13 41 20·73	60·33	2	3·454	+ 0·0293	− 0·002
556	4607	2	85 Ursæ Majoris η	13 42 1·03	56·43	1	2·386	− 0·0105	− 0·012
557	4608	5	89 Virginis.........	13 42 16·21	59·29	10	3·253	+ 0·0163	− 0·009
558	4638	3	Centauri.......... ζ	13 46 49·43	60·38	2	3·707	+ 0·0468	− 0·012
559	4648	3	8 Boötis............ η	13 48 1	..	..	2·862	− 0·0007	− 0·004
560	4660	5	Apodis............. θ	13 51 48·88	56·46	4	+ 5·608	+ 0·2893	..

No.	Mean N P.D. 1860, Jan. 1.	Mean Year and Fraction of Year.	No. of Obs. of N.P.D.	Annual Precess. in N.P.D. for 1860.	Secular Variation of Precess.in N.P.D.	Annual Proper Motion in N.P.D.	No. for reference.			
							Lacaille.	Brisbane.	Fallows or Johnson.	Greenwich or Henderson.
	° ′ ″	1800		″	″	″				
526	132 37 16·73	60·38	3	+19·29	−0·144	− 0·04	5422	4353	..	..
527	105 26 31·70	60·33	1	19·26	−0·137	+ 0·30	..	4362	J 298	1042
528	157 9 3·47	60·35	2	19·23	−0·172	+ 0·04	5433	4369	..	..
529	113 8 28·38	56·11	5	19·13	−0·150	..	..	..	..	..
530	99 48 26·75	59·17	6	19·13	−0·146	+ 0·04	..	..	..	1052
531	175 5 48·76	56·47	8	19·02	−0·380	− 0·02	5452	4410	..	..
532	95 27 45·59	56·12	2	18·98	−0·155	+ 0·32	..	..	..	1060
533	94 11 26·64	56·21	2	18·96	−0·156	+ 0·02	..	..	..	1061
534	94 25 51·51	56·18	6	18·93	−0·158	+ 0·02	..	..	..	1062
535	100 25 45·48	58·27	118	18·91	−0·161	+ 0·04	..	4457	237·J 302	1063
536	175 3 52·49	56·64	28	18·88	−0·411	+ 0·04	5482	4445	..	..
537	105 14 47·18	59·29	1	18·85	−0·166	..	..	..	..	..
538	128 40 57·71	60·29	1	18·76	−0·185	+ 0·04	5569	4496	J 304	..
539	95 44 46·29	56·17	11	18·75	−0·170	− 0·02	..	..	..	1073
540	95 31 53·27	56·17	5	18·70	−0·172	+ 0·04	..	..	..	1076
541	118 50 36·27	60·41	2	18·70	−0·184	+ 0·03	5578	4519	..	1077
542	104 38 29·37	59·53	4	18·68	−0·178	+ 0·10	..	..	..	1078
543	99 26 31·75	58·45	9	18·67	−0·176	+ 0·03	..	..	..	1079
544	102 29 41·70	57·42	2	18·62	−0·180	+ 0·06	..	4542	..	1082
545	89 52 43·01	57·19	11	18·61	−0·175	− 0·06	..	..	..	1083
546	97 9 22·92	56·12	2	18·52	−0·184	+ 0·14	..	..	..	1087
547	142 45 8·97	60·49	1	18·49	−0·220	+ 0·02	5618	4570	239·J 305	H 45
548	97 59 42·14	56·37	1	18·38	−0·191	0·00	..	..	..	1096
549	105 28 23·14	58·54	1	18·29	−0·201	+ 0·06	..	4616	..	1099
550	122 20 2·67	60·37	3	18·26	−0·214	+ 0·22	5668	4619	J 306	1102
551	105 3 44·80	58·28	4	18·25	−0·203	+ 0·11	..	4623	..	1103
552	101 43 24·05	56·15	1	18·23	−0·202	− 0·01	..	..	..	1104
553	115 57 24·80	56·12	5	18·13	−0·216	..	..	..	..	..
554	131 46 26·39	60·47	1	18·13	−0·231	− 0·02	5684	4645	J 308	..
555	123 45 0·90	60·31	1	18·13	−0·223	+ 0·17	5688	4647	J 309	1108
556	39 59 8·77	56·44	2	18·10	−0·158	+ 0·03	..	..	..	1109
557	107 26 6·61	59·04	12	18·09	−0·213	+ 0·03	..	4653	..	1112
558	136 35 49·21	60·38	2	17·91	−0·251	+ 0·05	5737	4683	240·J 312	H 57
559	70 53 55·96	56·55	1	17·87	−0·197	+ 0·36	. ..	..	..	1121
560	166 7 4·13	56·46	4	+17·72	−0·391	− 0·01	5757	4712	..	..

No.	No. in B.A.C.	Magnitude	Star's Name.	Mean R.A. 1860, Jan. 1.	Mean Year and Fraction of Year.	No. of Obs. of R.A.	Annual Precess. in R.A. for 1860.	Secular Variation of Precess. in R.A.	Annual Proper Motion in R.A.
				b m s	1800				
561	4668	.5	Centauri......... ν²	13 53 0·48	60·33	1	+3·703	+0·0440	-0·010
562	4469	1	Centauri......... β	13 53 58·60	60·33	10	4·163	+0·0837	-0·010
563	4672	4·5	93 Virginis........ r	13 54 31·48	60·39	4	3·047	+0·0064	+0·001
564	4681	5	Centauri..,....... χ	13 57 30·76	60·48	2	3·632	+0·0375	-0·011
565	4686	3	5 Centauri.........θ	13 58 27	..	..	3·546	+0·0317	-0·051
566	4700	5·6	Piazzi XIII. 317..	14 3 12·00	58·48	15	3·264	+0·0156	-0·005
567	4705	5	Octantis..,........ δ	14 4 56·05	56·44	5	8·734	+0·9831	-0·063
568	4716	4·5	98 Virginis........ κ	14 5 25	..	..	3·190	+0·0123	+0·001
569	4712	5	Apodis..,......... ι	14 5 40·19	56·43	4	6·817	+0·4771	-0·005
570	4722	6	Piazzi XIV. 22.....	14 7 41·49	58·73	10	3·296	+0·0168	+0·008
571	4729	1	16 Boötis......... a	14 9 16·60	58·45	14	2·813	+0·0003	-0·079
572	4730	6	Centauri............	14 9 37·59	57·46	1	4·345	+0·0889	..
573	4743	5·4	100 Virginis...... λ	14 11 32·26	60·33	1	3·236	+0·0140	-0·002
574	4768	5	Lupi............. r¹	14 17 10·11	60·57	1	3·815	+0·0437	-0·004
575	4784	6·5	52 Hydræ...........	14 19 58·92	60·42	2	3·495	+0·0250	+0·003
576	4790	6·7	Octantis.......... Z	14 23 43·65	58·79	81	21·523	+7·4007	-0·168
577	4808	4·3	25 Boötis......... ρ	14 25 47	..	..	2·595	-0·0016	-0·008
578	4811	.3	Centauri.......... η	14 26 37·82	60·48	1	3·779	+0·0388	-0·009
579	4831	4	Centauri......... a¹	14 30 7·06	60·45	3	4·497	+0·0875	-0·470
580	4832	1	Centauri......... a²	14 30 7·23	58·85	2	4·497	+0·0875	-0·470
581	4852	5	Lacaille 6063.......	14 35 6·36	60·51	2	3·649	+0·0301	-0·008
582	4868	6	5 Libræ............	14 38 14·86	57·73	10	3·298	+0·0152	-0·003
583	4876	2·3	36 Boötis......... ε²	14 38 52	..	..	2·624	-0·0001	-0·005
584	4872	6	Centauri..........	14 38 54·06	57·46	1	4·342	+0·0700	-0·023
585	4880	5	56 Hydræ...........	14 39 34·87	60·21	1	3·481	+0·0220	0·000
586	4882	5	57 Hydræ...........	14 39 46·39	60·46	1	3·492	+0·0224	-0·005
587	4883	6	Octantis..	14 40 50·01	56·57	6	9·600	+0·9115	..
588	4895	2·3	9 Libræ........... a	14 43 8·35	58·21	18	3·314	+0·0155	-0·007
589	4896	6	Bradley 1895	14 43 45·52	57·27	2	3·343	+0·0164	-0·005
590	4913	6	12 Libræ............	14 46 12·57	58·47	1	3·468	+0·0207	+0·002
591	4916	5·6	Lacaille 6146.......	14 47 9·61	60·48	1	3·657	+0·0283	-0·004
592	4923	6	Piazzi XIV. 212.,..	14 49 17·77	58·54	1	3·414	+0·0185	+0·068
593	4924	3	Lupi .... β	14 49 22·53	60·41	2	3·899	+0·0392	-0·014
594	4928	.3	Centauri........... κ	14 50 4·13	60·47	2	3·873	+0·0378	-0·003
595	..	.7	Lacaille 6198........	14 54 25·83	60·47	3	+3·651	+0·0267	..

No.	Mean N.P.D. 1860, Jan. 1.	Mean Year and Fraction of Year.	No. of Obs. of N.P.D.	Annual Precess. in N.P.D.	Secular Variation of Precess.in N.P.D.	Annual Proper Motion in N.P.D.	No. for reference.			
							Lacaille.	Brisbane.	Fallows or Johnson.	Greenwich or Henderson.
	° ′ ″	1800		″	″	″				
561	134 55 21·85	60·33	1	+ 17·67	− 0·264	+ 0·00	5782	4729	··	··
562	149 41 42·38	57·52	11	17·63	− 0·297	+ 0·05	5784	4733	243.J 315	H 31
563	87 46 35·37	59·08	6	17·60	− 0·221	+ 0·07	··	··	··	1124
564	130 30 22·65	60·48	2	17·48	− 0·267	− 0·02	5810	4757	J 316	··
565	125 40 43·82	60·50	1	17·44	− 0·263	+ 0·64	5820	4766	245.J 317	1126
566	105 38 19·11	58·52	16	17·23	− 0·251	+ 0·06	··	4797	··	1131
567	173 1 14·70	56·44	5	17·15	− 0·668	+ 0·01	5802	4790	J 319	H 11
568	99 37 12·94	56·32	4	17·13	− 0·250	− 0·02	··	··	J 321	1136
569	169 27 29·67	56·94	5	17·12	− 0·526	+ 0·02	5828	4799	··	··
570	107 32 43·29	58·73	10	17·02	− 0·262	+ 0·01	··	··	··	1139
571	70 5 13·25	58·80	37	16·95	− 0·227	+ 1·93	··	4840	··	1141
572	150 37 14·53	57·46	1	16·93	− 0·345	··	5875	··	··	··
573	102 43 28·60	57·09	11·	16·84	− 0·264	− 0·02	··	··	246.J 324	1143
574	134 35 8·62	60·51	1	16·57	− 0·320	+ 0·09	5928	4902	J 328	··
575	118 51 35·73	60·42	2	16·43	− 0·299	+ 0·04	5949	4925	··	1149
576	177 33 54·65	58·68	78	16·24	− 1·849	+ 0·05	5823	4886	J 327	H 6
577	59 0 45·41	60·37	1	16·13	− 0·232	− 0·14	··	··	··	1153
578	131 32 24·23	60·48	1	16·09	− 0·336	− 0·01	5993	4968	248.J 332	H 66
579	150 15 18·68	60·45	3	15·91	− 0·406	− 0·83	6014	4990	J 335	H 29
580	150 15 19·55	58·56	160	15·91	− 0·406	− 0·83	6017	4991	249.J 336	H 30
581	124 34 4·34	60·51	2	15·64	− 0·340	+ 0·26	6063	5029	J 339	1163
582	104 52 0·52	57·73	10	15·46	− 0·313	+ 0·01	··	5055	··	1167
583	62 20 0·55	56·55	1	15·43	− 0·251	− 0·01	··	··	··	1170
584	146 4 29·26	57·46	1	15·43	− 0·412	+ 0·20	6082	5057	··	··
585	115 29 53·47	60·21	3	15·39	− 0·332	+ 0·03	6102	5060	··	1172
586	116 3 24·60	60·46	1	15·38	− 0·334	+ 0·02	6104	5061	··	1173
587	172 28 8·84	58·58	8	15·32	− 0·910	+ 0·05	··	5046	··	··
588	105 27 26·90	58·17	92	15·19	− 0·322	+ 0·06	··	··	251.J 343	1177
589	107 12 18·50	57·89	3	15·15	− 0·326	+ 0·17	··	··	··	1178
590	114 4 0	··	··	15·01	− 0·342	+ 0·03	6143	··	··	1181
591	123 17 2·99	60·48	1	14·95	− 0·362	+ 0·04	6146	5115	··	1183
592	110 46 51·76	58·54	1	14·83	− 0·342	+ 1·68	··	··	··	1186
593	132 33 59·60	60·41	2	14·82	− 0·389	+ 0·03	6160	5129	J 344	H 64
594	131 32 20·49	60·47	2	14·78	− 0·388	+ 0·01	6170	5133	'J 346	H 65
595	122 5 16·96	60·47	3	+ 14·52	− 0·373	··	6198	··	··	··

No.	No. in B.A.C.	Magnitude	Star's Name	Mean R.A. 1860, Jan. 1.	Mean Year and Fraction of Year.	No. of Obs. of R.A.	Annual Precess. in R.A. for 1860.	Secular Variation of Precess. in R.A.	Annual Proper Motion in R.A.
				h m s	1800				
596	4947	7	Piazzi XIV. 246...	14 55 14·66	58·42	5	+3·356	+0·0161	+0·011
597	4948	5	Lupi .... π	14 55 36·09	60·45	2	4·049	+0·0451	−0·009
598	4950	3·4	20 Libræ.............	14 55 53·04	58·62	3	3·500	+0·0209	−0·007
599	4969	4·5	43 Boötis .... ψ	14 58 26·77	58·44	4	2·583	+0·0011	−0·013
600	..	7	Lacaille 6229........	14 58 37·60	60·56	1	3·668	+0·0268	..
601	4970	6	21 Libræ .... ν'	14 58 49·32	57·00	3	3·337	+0·0153	−0·004
602	4973	5	Lupi .... λ	14 59 25·75	60·37	1	4·007	+0·0418	−0·011
603	4986	5	Lupi................. κ	15 2 13·09	60·47	2	4·140	+0·0475	−0·020
604	4995	5·4	24 Libræ......... ι'	15 4 14·82	59·53	13	3·408	+0·0171	−0·002
605	4997	7	B.F. 2065............	15 4 15	..	..	3·396	+0·0167	..
606	5005	3	Trianguli Australis γ	15 5 53·72	60·48	1	5·491	+0·1395	−0·018
607	5028	5	Lupi .... μ	15 8 48·55	60·36	1	4·137	+0·0452	−0·015
608	5029	7	Lupi...................	15 8 50	..	..	4·138	+0·0452	..
609	5032	4·5	2 Lupi	15 9 19·16	60·46	1	3·631	+0·0238	−0·004
610	5034	2	27 Libræ......... β	15 9 28·64	59·24	13	3·225	+0·0118	−0·009
611	5037	6	Octantis............ ρ	15 11 37·51	58·77	36	12·488	+1·3640	+0·078
612	5046	4	Lupi.............. δ	15 12 11·77	60·47	1	3·911	+0·0340	+0·001
613	5055	6	28 Libræ.............	15 12 57·73	58·42	5	3·389	+0·0159	0·000
614	5060	5	Lupi............. φ²	15 14 13·43	60·41	1	3·810	+0·0296	−0·004
615	5089	4	32 Libræ .... ζ'	15 20 21	..	..	3·371	+0·0148	+0·002
616	5087	7	Normæ...............	15 20 23·70	57·57	1	4·428	+0·0554	..
617	..	8	....	15 20 34·75	59·25	5	3·074	+0·0083	..
618	5100	6	34 Libræ .... ζ'	15 22 45	..	..	3·372	+0·0147	+0·004
619	5104	7	Piazzi XV. 91......	15 23 40·52	58·36	2	3·443	+0·0165	0·000
620	5103	5	Trianguli Australis ε	15 23 57·29	60·46	9	5·388	+0·1124	−0·002
621	5107	6	Apodis ....-	15 24 49·60	37·69	6	7·135	+0·2725	..
622	5112	6	35 Libræ .... ζ⁴	15 25 0	..	..	3·378	+0·0147	−0·001
623	5118	3	Lupi .... γ	15 25 49·41	60·48	3	3·971	+0·0332	−0·005
624	5121	6	36 Libræ.............	15 26 8·60	60·12	1	3·619	+0·0211	−0·004
625	5138	4·5	39 Libræ.............	15 28 31·80	59·52	2	3·625	+0·0210	−0·002
626	5139	5	Lupi .... ε	15 28 37·75	60·48	1	4·026	+0·0346	−0·039
627	5143	2	5 Coronæ Borealis α	15 28 45·61	57·44	4	2·529	+0·0024	+0·009
628	5144	7	Normæ...............	15 29 12·09	57·56	1	4·667	+0·0637	..
629	5151	4·5	40 Libræ.............	15 30 3·96	60·26	2	3·668	+0·0221	−0·007
630	5159	6·7	Normæ...............	15 30 58·69	57·57	1	+4·481	+0·0534	..

621. The R.A. has been brought up with Precession alone.

No.	Mean N.P.D. 1860, Jan. 1.	Mean Year and Fraction of Year.	No. of Obs. of N.P.D.	Annual Precess. in N.P.D. for 1860.	Secular Variation of Precess. in N.P.D.	Annual Proper Motion in N.P.D.	No. for reference.			
							Lacaille.	Brisbane.	Fallows or Johnson.	Greenwich or Henderson.
	° ′ ″	1800		′	″	″ ′				
596	107 4 41·67	58·42	5	+ 14·47	− 0·345	+ 0·09	..	..	..	1191
597	136 29 59·63	60·45	2	14·45	− 0·416	+ 0·04	6201	5166	J 348	..
598	114 43 44·04	58·66	15	14·43	− 0·361	+ 0·03	6212	5169	252·J 349	1194
599	62 30 15·51	58·44	4	14·28	− 0·271	0·00	..	....	.. ..	1196
600	122 21 55·19	60·56	1	14·26	− 0·382	..	6229	..	..	..
601	105 42 40·46	57·00	3	14·25	− 0·348	+ 0·03	..	..	..	1197
602	134 44 14·90	60·37	1	14·22	− 0·418	− 0·03	6232	5185	J 350	..
603	138 12 6·45	60·47	2	14·04	− 0·437	+ 0·06	6246	5205	J 352	..
604	109 15 32·49	59·53	13	13·92	− 0·364	+ 0·04	..	..	..	1205
605	108 34 27·58	58·45	1	13·92	− 0·363	..	..	..	..	1206
606	158 9 27·77	60·48	1	13·81	− 0·587	+ 0·03	6255	5227	253·J 353	H 18
607	137 21 23·23	58·44	2	13·63	− 0·449	+ 0·08	6296	5260	J 355	..
608	137 21 37·99	56·52	1	13·62	− 0·449	+ 0·06	..	5261	..	..
609	119 37 48·78	60·46	1	13·59	− 0·394	+ 0·08	6304	5266	J 356	..
610	98 51 49·32	58·65	42	13·58	− 0·352	+ 0·01	..	5270	254·J 357	1215
611	173 59 11·71	58·80	37	13·45	− 1·357	− 0·03	6216	5240	..	..
612	130 8 12·47	60·47	1	13·41	− 0·430	− 0·03	6326	5285	J 358	..
613	107 38 50·02	58·42	5	13·36	− 0·375	+ 0·08	..	..	..	1221
614	126 21 12·08	60·41	1	13·28	− 0·422	+ 0·07	6349	5299	..	1223
615	106 13 30·34	56·15	1	12·87	− 0·383	+ 0·05	..	..	..	1226
616	142 53 13·49	57·57	1	12·87	− 0·499	+ 0·01	6383	5345	..	..
617	90 7 24·23	59·25	5	12·86	− 0·350	..	..	..	..	..
618	106 7 34·04	56·51	1	12·71	− 0·385	+ 0·03	..	..	..	1744*
619	109 40 58·40	58·36	2	12·64	− 0·395	+ 0·09	..	..	..	1229
620	155 50 27·73	60·46	9	12·62	− 0·616	+ 0·10	6398	5372	J 362	..
621	165 36 53·96	56·52	1	12·57	− 0·816	+ 0·05	6381	5368	..	..
622	106 22 30·13	56·50	1	12·55	− 0·388	+ 0·02	..	..	..	..
623	130 41 32·46	60·48	3	12·50	− 0·458	+ 0·02	6422	5380	255·J 363	H 68
624	117 34 19·60	60·12	1	12·48	− 0·418	+ 0·09	6430	5385	..	..
625	117 40 4·60	59·39	3	12·31	− 0·423	0·00	6445	5400	J 366	1237
626	132 6 14·77	60·48	1	12·30	− 0·469	− 0·09	6443	5399	..	..
627	62 48 42·72	57·37	9	12·30	− 0·297	+ 0·07	..	..	..	1238
628	146 27 6·38	57·56	1	12·26	− 0·543	+ 0·03	6440	5401	..	..
629	119 18 49·99	60·26	2	12·21	− 0·430	+ 0·08	6455	5406	J 367	1239
630	142 55 57·44	57·57	1	+ 12·14	− 0·525	− 0·07	6451	5408	..	..

No	No. in B.A.C.	Magnitude	Star's Name.	Mean R.A. 1860, Jan. 1.	Mean Year and Fraction of Year.	No. of Obs. of R.A.	Annual Precess. in R.A. for 1860.	Secular Variation of Precess. in R.A.	Annual Proper Motion in R.A.
				h m s	1800				
631	5165	.5	Lacaille 6464.......	15 31 34.77	60.49	2	+ 4.109	+ 0.0370	− 0.018
632	5166	5.6	42 Libræ.............	15 32 0.78	58.19	5	3.533	+ 0.0180	− 0.003
633	5176	5	43 Libræ........... κ	15 33 53.16	59.33	2	3.447	+ 0.0157	− 0.003
634	..	.11	....	15 33 55.26	59.29	4	3.138	+ 0.0091	..
635	..	.8	....	15 35 20.41	60.47	5	3.725	+ 0.0230	..
636	5190	.6	44 Libræ........ η	15 36 12.21	56.28	3	3.367	+ 0.0136	+ 0.001
637	5196	2.3	24 Serpentis ...... a	15 37 22.42	59.26	7	2.941	+ 0.0062	+ 0.009
638	5197	6	Lalande 28670......	15 37 30.21	57.35	1	3.562	+ 0.0182	..
639	5200	.7	Normæ	15 38 3.47	57.57	1	4.568	+ 0.0546	..
640	5209	.6	Normæ.............	15 39 31.48	57.56	1	4.511	+ 0.0513	+ 0.013
641	5217	6	Trianguli Australis	15 40 11	..	..	5.393	+ 0.0982	..
642	5224	5	Trianguli Australis κ	15 41 42.63	60.57	2	5.822	+ 0.1249	− 0.008
643	5227	5	5 Lupi.. .... χ	15 42 4.21	60.50	4	3.793	+ 0.0239	− 0.001
644	5232	5	1 Scorpii........... b	15 42 33.89	60.16	5	3.594	+ 0.0184	− 0.005
645	5233	3	Trianguli Australis β	15 42 50.62	59.12	3	5.239	+ 0.0866	− 0.027
646	5251	.6	45 Libræ.......... λ	15 45 12.75	57.99	5	3.471	+ 0.0152	− 0.001
647	5250	5	2 Scorpii......... A²	15 45 12.86	57.28	2	3.589	+ 0.0180	− 0.003
648	5263	6	Normæ..............	15 46 59.22	57.52	2	4.303	+ 0.0399	..
649	5268	4.5	{ Lupi.. .... ξ¹	15 47 56.80	60.41	5	3.816	+ 0.0204	+ 0.005
650	5269	..	{ Lupi.............. ξ²	15 47 57.50	60.40	3	3.816	+ 0.0204	− 0.023
651	5272	5.4	5 Scorpii.......... ρ	15 48 14.84	59.58	12	3.689	+ 0.0201	− 0.004
652	5277	6	Normæ.....	15 49 6.04	57.56	1	4.598	+ 0.0508	+ 0.019
653	5289	.3	6 Scorpii,......... π	15 50 23.41	57.73	6	3.616	+ 0.0180	− 0.003
654	5292	4.5	Lupi................. η	15 50 51.08	60.45	2	3.955	+ 0.0270	− 0.011
655	5303	2.3	7 Scorpii........... δ	15 52 3.55	56.68	1	3.535	+ 0.0160	− 0.001
656	5309	.6	Bradley 2031	15 53 54	..	..	2.976	+ 0.0064	− 0.002
657	..	.10	....	15 55 7.69	59.27	7	3.278	+ 0.0108	..
658	5320	6	Normæ..............	15 56 12.90	57.57	1	4.760	+ 0.0547	..
659	5323	5	Normæ............. δ	15 56 36.63	36.23	8	4.212	+ 0.0336	..
660	5329	2	8 Scorpii.......... β¹	15 57 18.10	60.00	26	3.477	+ 0.0142	− 0.002
661	5331	4.5	Lupi............,...... θ	15 57 24.44	60.42	1	3.921	+ 0.0247	+ 0.004
662	5337	.4	9 Scorpii ...... ω¹	15 58 37.	..	..	3.499	+ 0.0145	+ 0.001
663	5339	5.6	Apodis............. δ¹	15 59 34.34	36.61	7	8.674	+ 0.3424	..
664	5347	5	Lacaille 6702.......	15 59 35.82	60.43	2	3.635	+ 0.0173	+ 0.010
665	5340	.6	Apodis............. δ²	15 59 41	..	..	+ 8.662	+ 0.3408	..

659. The R.A. has been brought up with Precession alone.
663. The R.A. has been brought up with Precession alone.

No.	Mean N.P.D. 1860, Jan. 1.	Mean Year and Fraction of Year.	No. of Obs. of N.P.D.	Annual Precess. in N.P.D. for 1860.	Secular Variation of Precess. in N.P.D.	Annual Proper Motion in N.P.D.	No. for reference.			
							Lacaille.	Brisbane.	Fallows or Johnson.	Greenwich or Henderson.
	° ′ ″	1800		″	″	″				
631	134 11 36·75	60·49	2	+ 12·10	− 0·483	+ 0·24	6464	5416	..	..
632	113 21 35·15	58·01	6	12·07	− 0·417	+ 0·02	6479	5423	..	1241
633	109 13 17·77	59·33	2	11·94	− 0·409	+ 0·12	..	..	J 368	1243
634	93 32 23·66	59·29	3	11·94	− 0·373	..	..	..	..	..
635	121 9 7·18	60·47	5	11·83	− 0·444	..	....	..	..	..
636	105 13 24·39	57·36	4	11·77	− c·403	+ 0·06	...	..	J 369	1249
637	83 7 52·06	58·24	17	11·69	− 0·354	− 0·05	..	5473	..	1251
638	114 16 19·07	56·93	2	11·68	− 0·427	− 0·14	..	5464	..	1252
639	143 57 28·08	57·57	1	11·64	− 0·547	− 0·07	..	5465	..	..
640	142 46 29·13	57·56	1	11·54	− 0·543	+ 0·06	6520	5484	..	..
641	154 43 26·94	56·52	1	11·49	− 0·648	..	6507	..	..	..
642	158 10 47·10	60·57	2	11·38	− 0·704	+ 0·04	6518	5491	..	..
643	123 11 49·53	60·50	4	11·36	− 0·461	+ 0·06	6548	5499	J 370	1256
644	115 19 19·58	59·99	6	11·32	− 0·438	+ 0·02	6557	..	J 373	1257
645	152 59 37·32	59·60	6	11·30	− 0·636	+ 0·43	6533	5497	256.J 371	H 23
646	109 44 42·45	57·99	5	11·13	− 0·426	+ 0·01	..	..	J 375	1263
647	114 54 20·13	57·28	2	11·13	− 0·440	+ 0·01	6574	5521	J 374	1264
648	137 44 42·46	57·52	2	11·00	− 0·530	+ 0·01	6580	5529	..	..
649	123 33 9·00	60·41	5	10·93	− 0·471	+ 0·01	6592	5535	..	1271
650	123 33 2·00	60·40	3	10·93	− 0·471	− 0·02	..	..	..	1272
651	118 48 5·74	59·58	12	10·91	− 0·456	+ 0·03	6601	5538	J 377	1273
652	143 36 52·36	57·56	1	10·84	− 0·569	+ 0·06	6589	5542	..	..
653	115 42 27··03·79	57·73	6	10·75	− 0·450	+ 0·04	6622	..	J 378	1279
654	127 59 33·32	60·45	2	10·71	− 0·493	+ 0·05	6619	5554	J 380	..
655	112 13 11·52	59·22	12	10·62	− 0·442	+ 0·01	..	5560	258.J 381	1281
656	85 10 41·14	56·50	1	10·49	− 0·374	..	..	..	..	..
657	100 14 18·63	59·27	5	10·40	− 0·413	..	..	..	..	..
658	145 48 22·82	57·57	1	10·31	− 0·599	+ 0·07	6650	5577	..	..
659	134 47 20·04	60·37	1	10·28	− 0·532	− 0·02	6664	5581	J 382	..
660	109 25 8·04	58·92	77	10·23	− 0·441	+ 0·02	..	..	260.J 385	1285
661	126 25 0·15	60·42	1	10·22	− 0·496	+ 0·08	6678	5591	J 384	1287
662	110 17 11·52	58·33	1	10·13	− 0·445	+ 0·01	..	..	J 386	1289
663	168 20 3·15	56·52	2	10·06	− 1·099	0·00	6623	5584	..	..
664	115 56 52·23	60·43	2	10·06	− 0·463	+ 0·07	6702	5605	..	1291
665	168 18 23·65	56·52	2	+ 10·05	− 1·097	+ 0·03	6628.	5586 .	... ..	..

No.	No. in B.A.C.	Magnitude	Star's Name,	Mean R.A. 1860, Jan. 1.	Mean Year and Fraction of Year.	No. of Obs. of R.A.	Annual Precess. in R.A. for 1860.	Secular Variation of Precess. in R.A.	Annual Proper Motion in R.A.
				h m s	1800		s	s	s
666	5374	6	Lacaille 6725........	16 2 20·57	60·47	2	+ 3·719	+ 0·0188	— 0·013
667	5382	4	14 Scorpii .... ν²	16 3 51·75	60·19	1	3·477	+ 0·0136	— 0·002
668	5384	6	Normæ...............	16 4 16·97	38·02	5	4·912	+ 0·0568	..
669	5395	6	Piazzi XVI. 10.....	16 5 27·04	58·31	1	3·523	+ 0·0143	— 0·002
670	..	9	....	16 6 18·83	59·24	6	3·357	+ 0·0114	..
671	5414	3	1 Ophiuchi........ δ	16 7 0·72	60·09	9	3·141	+ 0·0082	— 0·006
672	5423	7	Piazzi XVI. 28.....	16 8 48·48	58·27	4	3·497	+ 0·0135	— 0·009
673	5425	5	Normæ............ γ²	16 9 22·77	60·34	8	4·474	+ 0·0380	— 0·018
674	5412	6·7	Brisbane 5607.......	16 9 40·31	58·44	22	20·372	+ 2·4765	+ 0·018
675	5435	6	Lacaille 6788........	16 10 41·71	60·43	1	3·774	+ 0·0187	+ 0·005
676	5436	7	Piazzi XVI. 39....	16 10 56·06	58·30	1	3·502	+ 0·0133	+ 0·007
677	5439	5	Apodis........ γ	16 12 6·60	56·53	3	8·970	+ 0·3291	— 0·045
678	..	7	Lacaille 6796........	16 12 34·23	60·55	1	3·844	+ 0·0199	..
679	5447	3·4	20 Scorpii........ σ	16 12 41·02	59·08	6	3·635	+ 0·0156	— 0·003
680	..	7·8	Lacaille 6815........	16 14 56·30	60·59	1	3·847	+ 0·0196	..
681	5467	5	4 Ophiuchi........ ψ	16 15 54·99	58·26	3	3·502	+ 0·0128	— 0·004
682	..	10	....	16 16 6·16	59·25	6	3·425	+ 0·0116	..
683	5485	6·7	Aræ................	16 18 5	..	..	4·962	+ 0·0510	..
684	5498	1·2	21 Scorpii........ α	16 20 49·69	59·63	41	3·667	+ 0·0151	— 0·001
685	5508	4	Normæ........... α	16 22 14·33	60·29	5	3·905	+ 0·0195	— 0·003
686	5516	5	8 Ophiuchi........ φ	16 23 7·78	59·25	7	3·429	+ 0·0110	— 0·001
687	5510	5	Apodis........... β	16 23 10·92	56·66	3	8·446	+ 0·2471	— 0·081
688	5536	5	Trian. Australis η¹	16 26 58·72	60·48	2	6·115	+ 0·0922	..
689	5538	5	Lacaille 6890........	16 27 10·04	60·32	6	3·931	+ 0·0191	— 0·003
690	5539	3·4	23 Scorpii........ τ	16 27 10·35	58·10	18	3·723	+ 0·0152	— 0·001
691	5565	6	Trian. Australis η²	16 32 30·58	40·57	2	6·118	+ 0·0865	..
692	5579	5	Bradley 2114.......	16 33 28·95	56·45	1	3·463	+ 0·0105	— 0·004
693	5578	2	Trian. Australis α	16 33 52·55	59 77	11	6·272	+ 0·0920	0·000
694	5588	6·7	Lacaille 6950........	16 34 38·99	60 ,7	5	3·845	+ 0·0161	..
695	..	9·10	....	16 34 39·58	59·28	8	3·562	+ 0·0118	..
696	5604	3·2	40 Herculis........ ζ	16 36 0	..	..	2·296	+ 0·0033	— 0·034
697	5609	4·5	Aræ................ η	16 37 42·81	59·60	3	5·138	+ 0·0459	0·000
698	5616	6·7	41 Herculis.........	16 38 10	..	..	2·931	+ 0·0052	— 0·015
699	5614	6	25 Scorpii............	16 38 17	..	..	3·663	+ 0·0127	+ 0·001
700	..	9	....	16 40 15·16	59·25	5	+ 3·572	+ 0·0112	..

668. The R.A. has been brought up with Precession alone.
691. The R.A. has been brought up with Precession alone.

No.	Mean N P.D. 1860, Jan. 1.	Mean Year and Fraction of Year.	No. of Obs. of N.P.D.	Annual Precess. in N.P.D. for 1860.	Secular Variation of Precess. in N.P.D.	Annual Proper Motion in N.P.D.	No. for reference.			
							Lacaille.	Brisbane.	Fallows or Johnson.	Greenwich or Henderson.
666	119 2 34·79	1800 60·47	2	+ 9·85	− 0·477	+ 0·10	6725	5629	..	..
667	109 5 36·46	60·19	1	9·73	− 0·448	+ 0·03	..	..	262.J 391	1295
668	147 33 3·34	56·51	1	9·70	− 0·630	+ 0·15	6722	5634	..	..
669	111 2 20·86	58·31	1	9·61	− 0·455	+ 0·05	..	..	..	1300
670	103 38 1·81	59·24	6	9·55	− 0·435	..	..	..	..	..
671	93 19 50·78	59·45	24	9·49	− 0·408	+ 0·13	..	..	263.J 393	1304
672	109 45 10·45	58·27	6	9·35	− 0·455	+ 0·16	..	..	..	1306
673	139 48 26·47	60·34	8	9·31	− 0·582	+ 0·06	6764	5675	J 396	..
674	176 4 52·81	58·44	22	9·29	− 2·637	0·00	..	5607	J 388	..
675	120 33 46·49	60·43	1	9·21	− 0·493	+ 0·09	6738	5685	..	1309
676	109 52 21·02	58·30	2	9·19	− 0·458	+ 0·12	..	..	..	1310
677	168 34 23·78	58·26	8	9·10	− 1·170	+ 0·07	6727	5678	J 397	..
678	122 5 44·30	60·55	1	.9·06	− 0·504	..	6796	..	..	..
679	115 15 11·28	59·38	14	9·05	− 0·477	− 0·01	6799	5703	J 399	1311
680	122 52 1·63	60·59	1	8·87	− 0·507	..	6815	..	..	..
681	109 42 21·71	58·26	4	8·80	− 0·463	+ 0·06	..	..	265.J 400	1318
682	106 22 31·64	59·26	6	8·79	− 0·453	..	..	..	..	..
683	147 26 21·21	56·52	1	8·63	− 0·657	+ 0·15	6827	5728	..	..
684	116 7 3·10	58·71	139	8·41	− 0·489	+ 0·03	6853	5743	267.J 404	1330
685	124 23 42·76	60·25	4	8·30	− 0·522	+ 0·08	6859	5747	J 405	1336
686	106 18 14·25	59·25	7	8·23	− 0·460	+ 0·03	..	..	268.J 406	1337
687	167 12 55·37	58 22	7	8·22	− 1·127	+ 0·31	6817	5742	..	..
688	158 0 37·63	60·48	2	7·91	− 0·823	0·00	6865	5756	..	..
689	124 57 45·75	60·32	6	7·90	− 0·530	+ 0·03	6890	5767	J 408	..
690	117 55 17·66	58·08	20	7·90	− 0·503	+ 0·02	6897	5768	269.J 409	1343
691	157 50 8·85	56·52	2	7·48	− 0·831	− 0·01	6900	5797	..	..
692	107 28 2·89	56·45	1	7·39	− 0·473	− 0·01	..	..	271.J 413	..
693	158 45 49·57	59·44	21	7·36	− 0·854	+ 0·06	6911	5804	270.J 412	H 17
694	121 50 8·64	60·47	5	7·30	− 0·526	..	6950	..	..	1352
695	111 29 58·16	59·28	5	7·30	− 0·487	..	..	..	..	..
696	58 8 29·25	59·33	1	7·19	− 0·316	− 0·45	..	..	..	1355
697	148 47 7·45	60·01	5	7·05	− 0·705	+ 0·02	6956	5828	J 414	..
698	83 38 23·78	56·52	1	7·01	− 0·404	+ 0·16	..	..	..	1356
699	115 16 10·81	56·50	1	7·00	− 0·504	..	6981	..	..	..
700	111 41 17·76	59·26	4	+ 6·84	− 0·493	..	..	..	..	..

No.	No. in B.A.C.	Magnitude.	Star's Name.	Mean R.A. 1860, Jan. 1.	Mean Year and Fraction of Year.	No. of Obs. of R.A.	Annual Precess. in R.A. for 1860.	Secular Variation of Precess. in R.A.	Annual Proper Motion in R.A.
				h m s	1800		"	"	"
701	5632	3	26 Scorpii......... ε	16 41 6·19	60·54	4	+ 3·922	+ 0·0162	− 0·051
702	..	6	....	16 41 14·19	59·27	6	3·570	+ 0·0111	..
703	5638	3	Scorpii.......... μ¹	16 42 23·49	60·48	3	4·050	+ 0·0182	− 0·007
704	5651	4·5	Scorpii.......... ζ¹	16 44 7·68	60·36	1	4·216	+ 0·0207	+ 0·003
705	5661	3	Scorpii.......... ζ²	16 44 44·29	60·25	3	4·216	+ 0·0206	− 0·020
706	5683	3·4	Aræ............... ζ	16 47 2·69	60·52	5	4·938	+ 0·0354	− 0·013
707	5697	4	Aræ............... ε¹	16 48 26·16	60·49	3	4·757	+ 0·0307	− 0·008
708	5708	3·4	27 Ophiuchi...... κ	16 51 2·53	59·08	6	2·856	+ 0·0044	− 0·023
709	5711	6	26 Ophiuchi........	16 51 35	..	..	3·662	+ 0·0109	+ 0·001
710	5713	5	Aræ............... ε²	16 51 58·23	60·12	5	4·770	+ 0·0295	− 0·006
711	..	8	....	16 53 10·88	59·25	3	3·745	+ 0·0117	..
712	5723	6	29 Ophiuchi........	16 53 (40·02)	..	..	3·505	+ 0·0089	− 0·001
713	5724	6	30 Ophiuchi........	16 53 (40·35)	..	..	3·162	+ 0·0060	+ 0·001
714	..	7·8	....	16 55 12·14	59·26	4	3·756	+ 0·0115	..
715	5735	5	Lacaille 7109......	16 55 36·98	60·41	1	3·936	+ 0·0137	− 0·007
716	..	8·9	....	16 57 52·22	59·27	3	3·759	+ 0·0111	..
717	5760	6	Piazzi XVI. 289...	16 58 19	..	..	3·088	+ 0·0054	..
718	..	10	....	16 58 46·57	59·36	4	3·756	+ 0·0109	..
719	..	8	....	16 59 32·34	59·50	4	3·756	+ 0·0108	..
720	5778	3·4	Scorpii............ η	17 2 7·86	60·47	16	4·281	+ 0·0170	− 0·003
721	5781	2·3	35 Ophiuchi...... η	17 2 21	..	..	3·432	+ 0·0074	+ 0·001
722	5789	7·8	Ophiuchi............	17 3 28	..	..	3·729	+ 0·0100	..
723	..	10	....	17 5 18·09	59·27	5	3·818	+ 0·0106	..
724	5794	6	Lacaille 7088........	17 5 22·52	56·67	2	11·019	+ 0·2678	+ 0·011
725	..	9	....	17 5 45·14	59·31	4	3·813	+ 0·0107	..
726	5803	5·6	Apodis............ ι	17 6 30·44	56·66	2	6·640	+ 0·0661	+ 0·007
727	5808	5	36 Ophiuchi...... A	17 6 44·60	56·86	8	3·718	+ 0·0094	− 0·037
728	5806	6	Aræ................	17 6 50·16	57·56	1	5·284	+ 0·0325	+ 0·015
729	5809	6·7	Lacaille 7191........	17 6 55·15	59·36	4	3·824	+ 0·0104	..
730	5817	5·6	Lacaille 7202........	17 7 57·44	60·40	5	3·902	+ 0·0110	− 0·005
731	5821	Var.	64 Herculis...... a¹	17 8 15·91	58·44	2	2·734	+ 0·0035	− 0·003
732	5820	7	Lacaille 7212......	17 8 24·33	59·44	4	3·824	+ 0·0102	..
733	5826	7	Lacaille 7222........	17 9 27·85	59·43	4	3·817	+ 0·0099	..
734	5836	6	Aræ................	17 10 51·89	57·57	1	5·152	+ 0·0275	..
735	..	7	....	17 12 22·63	59·37	4	+ 3·871	+ 0·0100	..

No.	Mean N.P.D. 1860. Jan. 1.	Mean Year and Fraction of Year.	No. of Obs. of N.P.D.	Annual Precess. in N.P.D. for 1860.	Secular Variation of Precess. in N.P.D.	Annual Proper Motion in N.P.D.	No. for reference.			
							Lacaille.	Brisbane.	Fallows or Johnson.	Greenwich or Henderson
	° ′ ″	1800		″	″	″				
701	124 2 6·25	60·54	4	+ 6·77	− 0·541	+ 0·33	6996	5851	272.J 415	1359
702	111 36 8·30	59·28	7	6·76	− 0·493	··	··	··	··	··
703	127 48 9·19	60·48	3	6·66	− 0·560	0·00	7006	5860	J 416	H 73
704	132 7 26·31	60·36	1	6·52	− 0·586	− 0·04	7016	5873	··	··
705	132 7 0·45	60·25	3	6·47	− 0·585	+ 0·20	7025	5881	··	··
706	145 45 48·75	59·74	5	6·28	− 0·687	+ 0·08	7034	5892	J 419	··
707	142 56 21·99	60·49	3	6·16	− 0·663	0·00	7050	5900	J 421	··
708	80 24 15·10	59·07	16	5·94	− 0·400	− 0·02	··	··	··	1371
709	114 46 15·11	57·20	1	5·90	− 0·513	+ 0·14	7085	··	··	1373
710	143 1 17·41	59·67	7	5·87	− 0·668	+ 0·16	7073	5921	··	··
711	117 43 42·99	59·25	4	5·77	− 0·526	··	··	··	··	··
712	108 40 33·00	56·50	1	5·72	− 0·493	− 0·01	··	··	··	1917*
713	94 0 31·24	60·62	1	5·72	− 0·443	+ 0·08	··	··	··	··
714	118 3 9·52	59·26	4	5·60	− 0·528	··	··	··	··	··
715	123 55 17·27	60·41	1	5·56	− 0·554	+ 0·05	7109	5950	J 422	1379
716	118 4 5·37	59·27	4	5·38	− 0·530	··	··	··	··	··
717	90 41 47·67	58·56	2	5·33	− 0·436	··	··	··	··	··
718	117 56 4·93	59·34	5	5·30	− 0·530	··	··	··	··	··
719	117 54 49·89	59·50	4	5·23	− 0·531	··	··	··	··	··
720	133 2 58·80	60·24	17	5·01	− 0·607	+ 0·26	7155	5987	J 424	H 61
721	105 32 49·86	60·62	2	4·99	− 0·487	− 0·12	··	··	276.J 425	1384
722	116 51 47·80	56·52	1	4·90	− 0·529	··	7165	··	··	··
723	119 52 44·65	59·27	5	4·75	− 0·543	··	··	··	··	··
724	170 43 0·18	58·26	6	4·74	− 1·563	+ 0·10	7088	5982	··	··
725	119 41 24·77	59·31	4	4·71	− 0·543	··	··	··	··	··
726	159 58 6·04	57·61	5	4·64	− 0·944	+ 0·06	7156	5999	··	··
727	116 23 33·28	56·90	9	4·62	− 0·530	+ 1·12	7192	··	J 426	1390
728	149 32 8·63	57·56	1	4·61	− 0·752	+ 0·11	7170	6006	··	··
729	120 2 41·08	59·36	4	4·61	− 0·545	··	7191	··	··	··
730	122 30 1·56	60·40	5	4·51	− 0·556	− 0·10	7202	··	··	1394
731	75 26 49·41	59·21	5	4·49	− 0·391	− 0·04	··	6026	··	1395
732	120 0 17·71	59·44	4	4·48	− 0·546	··	7212	··	··	··
733	119 43 3·05	59·43	4	4·39	− 0·545	··	7222	··	··	··
734	147 51 49·91	57·57	1	4·27	− 0·736	− 0·05	7213	6035	··	··
735	121 26 5·25	59·38	4	+ 4·14	− 0·554	··	··	··	··	··

No.	No. in B.A.C.	Magnitude,	Star's Name.	Mean R.A. 1860, Jan. 1.	Mean year and Fraction of year.	No. of Obs. of R.A.	Annual Precess. in R.A. for 1860.	Secular Variation of Precess. in R.A.	Annual Proper Motion in R.A.
				h m s	1800				
736	..	7·8	....	17 13 12·77	59·51	4	+ 3·872	+ 0·0099	..
737	5851	3·4	42 Ophiuchi ...... θ	17 13 24·84	59·64	23	3·679	+ 0·0081	— 0·003
738	5850	3	Aræ............ γ	17 13 37·09	60·57	3	5·031	+ 0·0241	— 0·004
739	5852	3	Aræ............ β	17 13 40·30	60·38	1	4·970	+ 0·0231	+ 0·002
740	5855	6	Scorpii............	17 14 4	..	..	4·338	+ 0·0146	..
741	5857	6	43 Ophiuchi........	17 14 33·10	57·36	1	3·769	+ 0·0087	+ 0·003
742	5859	5	Aræ............ κ¹	17 15 5·47	60·14	3	4·663	+ 0·0182	..
743	..	7·8	....	17 17 25·47	59·29	5	3·920	+ 0·0093	..
744	5876	5	44 Ophiuchi...... b	17 17 49·42	59·52	3	3·658	+ 0·0074	— 0·002
745	5881	5	45 Ophiuchi...... d	17 18 25·04	59·60	16	3·823	+ 0·0085	— 0·002
746	5877	4	Aræ............ δ	17 18 28·45	34·54	5	5·402	+ 0·0273	..
747	5888	6	Serpentis............	17 19 10	..	..	3·361	+ 0·0055	..
748	5889	7	Aræ............	17 19 30·00	57·54	1	5·084	+ 0·0219	..
749	..	7·8	....	17 19 52·28	59·29	6	3·923	+ 0·0090	..
750	5899	3	Aræ............ a	17 21 1·50	60·23	4	4·629	+ 0·0154	— 0·005
751	5901	3·4	34 Scorpii........ υ	17 21 14·90	60·52	5	4·072	+ 0·0100	— 0·007
752	..	6·7	Lacaille 7330......	17 23 8·41	59·32	4	3·968	+ 0·0087	..
753	..	9	....	17 24 2·91	59·26	4	3·972	+ 0·0085	..
754	5915	3	35 Scorpii .... λ	17 24 6·20	60·35	7	4·067	+ 0·0092	— 0·010
755	5925	5·6	Scorpii............	17 25 33·91	58·25	2	3·914	+ 0·0079	..
756	5930	6	Aræ............ π	17 26 36·58	57·54	1	4·920	+ 0·0163	..
757	5935	3	Scorpii .... θ	17 27 15·81	59·61	2	4·302	+ 0·0103	+ 0·001
758	5937	3·2	23 Draconis....... β	17 27 16·31	56·49	1	1·353	+ 0·0051	— 0·003
759	5934	6	Apodis,.............	17 27 31	..	..	7·181	+ 0·0500	..
760	5941	2	55 Ophiuchi ...... a	17 28 26·22	60·17	11	2·774	+ 0·0031	+ 0·004
761	..	9·10	....	17 29 24·52	59·46	5	4·012	+ 0·0078	..
762	5949	4·3	55 Serpentis........ ξ	17 29 34	..	..	3·435	+ 0·0048	— 0·004
763	5960	7	Lacaille 7382.......	17 30 53·86	60·49	2	3·905	+ 0·0068	+ 0·009
764	..	7	....	17 31 13·04	59·26	4	4·020	+ 0·0074	..
765	5964	7	Piazzi XVII. 167...	17 31 51·31	60·36	1	3·906	+ 0·0067	— 0·023
766	5936	6	Brisbane 6058.......	17 32 22·69	57·06	16	35·379	+ 1·9126	— 0·107
767	5970	3	Scorpii......... κ	17 32 48·37	60·31	3	4·145	+ 0·0078	0·000
768	5974	6	Lacaille 7397......	17 33 20·86	59·34	4	4·068	+ 0·0072	— 0·002
769	5976	5·4	56 Serpentis....... o	17 33 32	..	..	3·374	+ 0·0042	— 0·008
770	..	8	....	17 34 34·83	59·28	4	+ 4·062	+ 0·0070	..

746. The R.A. has been brought up with Precession alone.

No.	Mean N.P.D. 1860. Jan. 1.	Mean Year and Fraction of Year.	No. of Obs. of N.P.D.	Annual Precess. in N.P.D. for 1860.	Secular Variation of Precess. in N.P.D	Annual Proper Motion in N.P.D.	No. for reference.			
							Lacaille.	Brisbane.	Fallows or Johnson.	Greenwich or Henderson.
	o ′ ″	1800		″	″	″				
736	121 26 30·98	59·40	16	+ 4·07	− 0·555	··	··	··	··	··
737	114 51 20·39	58·95	53	4·05	− 0·527	− 0·02	7254	··	J 432	1405
738	146 14 24·34	60·55	4	4·03	− 0·720	0·00	7233	6048	278.J 429	H 39
739	145 23 30·16	56·52	1	4·03	− 0·712	+ 0·03	7237	6050	J 430	H 42
740	134 1 23·42	56·50	2	3·99	− 0·621	+ 0·10	7247	6051	··	··
741	118 0 9·33	57·36	1	3·95	− 0·541	0·00	7260	6059	··	1406
742	140 30 0·34	60·14	3	3·91	− 0·669	0·00	7253	6060	··	··
743	122 50 10·81	59·29	5	3·71	− 0·563	··	··	··	··	··
744	114 2 33·40	59·52	3	3·67	− 0·526	+ 0·12	7289	··	··	1407
745	119 44 9·44	58·87	29	3·62	− 0·550	+ 0·18	7293	··	J 435	1409
746	150 33 39·13	59·64	1	3·62	− 0·777	+ 0·13	7271	6081	J 434	··
747	102 23 5·56	56·53	1	3·55	− 0·484	··	··	··	··	··
748	146 48 12·97	57·54	1	3·53	− 0·731	··	7281	··	··	··
749	122 53 1·20	59·29	6	3·50	− 0·565	··	··	··	··	··
750	139 45 35·07	60·23	4	3·39	− 0·667	+ 0·10	7301	6094	279.J 436	H 49
751	127 10 45·78	60·52	5	3·38	− 0·587	+ 0·04	7313	6098	J 437	··
752	124 10 7·81	59·30	3	3·22	− 0·573	··	··	··	··	··
753	124 16 27·18	59·26	4	3·14	− 0·573	··	··	··	··	··
754	126 59 49·01	60·35	7	3·13	− 0·587	+ 0·02	7336	6116	280.J 439	H 75
755	122 28 48·95	58·25	2	3·00	− 0·566	··	··	6125	··	1414
756	144 24 3·18	56·85	3	2·91	− 0·711	+ 0·22	7342	6127	··	··
757	132 54 13·23	59·62	1	2·86	− 0·622	+ 0·02	7351	6134	281.J 440	H 62
758	37 35 27·27	56·49	1	2·85	− 0·197	0·00	··	··	··	1415
759	162 8 36·82	56·51	1	2·83	− 1·038	··	7317	··	··	··
760	77 20 5·96	59·76	20	2·75	− 0·402	+ 0·20	··	6145	··	1416
761	125 21 53·84	59·46	5	2·67	− 0·579	··	··	··	··	··
762	105 18 23·97	60·68	1	2·66	− 0·498	+ 0·04	··	··	282.J 441	1420
763	122 7 3·03	60·49	2	2·54	− 0·566	+ 0·06	7382	6156	··	1422
764	125 33 52·00	59·26	4	2·52	− 0·581	··	··	··	··	··
765	122 8 3·41	60·36	1	2·46	− 0·566	+ 0·07	··	6163	··	1424
766	177 38 51·45	57·77	25	2·41	− 5·124	+ 0·11	··	6058	J 433	··
767	128 57 10·48	59·36	4	2·37	− 0·601	+ 0·01	7393	6169	J 444	H 71
768	126 52 11·37	59·34	4	2·33	− 0·590	+ 0·10	7397	6174	··	·
769	102 47 47·85	60·43	1	2·31	− 0·490	+ 0·04	··	··	J 445	1426
770	126 42 6·70	59·28	4	+ 2·23	− 0·590	··	··	··	··	··

No.	No. in B.A.C.	Magnitude.	Star's Name.	Mean R.A. 1860, Jan. 1.	Mean Year and Fraction of Year.	No. of Obs. of R.A.	Annual Precess. in R.A. for 1860.	Secular Variation of Precess. in R.A.	Annual Proper Motion In R.A.
				h m s	1800				
771	5998	.6	Aræ............	17 36 58'33	57'54	1	+4'996	+ 0'0123	+ 0'016
772	6004	3'4	Scorpii.............. ι¹	17 37 47'67	60'19	3	4'191	+ 0'0068	— 0'003
773	6008	5	3 Sagittarii...........	17 38 44'97	59'34	8	3'773	+ 0'0049	— 0'010
774	6010	6	Pavonis...............	17 39 19'46	40'49	1	5'986	+ 0'0188	..
775	6016	.5'6	Lacaille 7451........	17 40 4'95	59'65	1	3'893	+ 0'0052	— 0'001
776	6018	4	Lacaille 7449.......	17 40 19'63	59'58	1	4'076	+ 0'0057	— 0'003
777	..	9'10	....	17 40 23'42	59'32	5	4'093	+ 0'0059	..
778	6019	5'6	Scorpii........... ι²	17 40 23'87	35'62	2	4'191	+ 0'0062	..
779	6021	3'4	86 Herculis....... μ	17 40 58'77	60'27	6	2'369	+ 0'0026	— 0'026
780	..	7	....	17 41 40'79	59'27	5	4'104	+ 0'0056	..
781	..	7	....	17 44 44'88	59'40	5	4'136	+ 0'0049	..
782	6049	6	Piazzi XVII. 265 .	17 45 16	..	..	3'328	+ 0'0031	— 0'003
783	..	7'8	....	17 46 5'42	59'46	5	4'139	+ 0'0047	..
784	5959	6	Octantis............ σ	17 48 7'95	58'05	27	109'007	+ 8'5420	+ 0'104
785	6074	5	Lacaille 7521.........	17 50 5'88	59'53	11	3'851	+ 0'0033	+ 0'003
786	..	7	Lacaille 7520........	17 50 35'92	59'27	4	4'163	+ 0'0037	..
787	..	8	....	17 50 47'51	59'26	4	4'179	+ 0'0037	..
788	6077	5	4 Sagittarii..........	17 51 14'75	59'60	2	3'661	+ 0'0028	— 0'005
789	6091	2'3	33 Draconis....... γ	17 53 21'25	56'49	1	1'391	+ 0'0031	0'000
790	..	7	Lacaille 7534......	17 54 46'75	59'34	4	4'219	+ 0'0027	..
791	6100	.5	Pavonis....... π	17 55 5'99	60'39	2	5'773	+ 0'0048	— 0'009
792	..	8	....	17 55 19'63	59'32	4	4'211	+ 0'0026	..
793	6104	5	69 Ophiuchi........ τ	17 55 27	..	..	3'264	+ 0'0022	+ 0'006
794	6105	4	Aræ.............. θ	17 55 43'92	60'63	2	4'671	+ 0'0028	— 0'007
795	6107	4	Sagittarii .... γ¹	17 56 4'64	58'48	4	3'831	+ 0'0022	+ 0'011
796	6112	6	Coronæ Australis...	17 56 42'60	37'92	.6	4'337	+ 0'0023	..
797	6115	3'4	10 Sagittarii....... γ²	17 56 48'95	59'57	4	3'857	+ 0'0021	— 0'004
798	6127	5	Piazzi XVII. 359...	17 59 12'91	58'63	1	3'797	+ 0'0017	— 0'001
799	6148	5	Lacaille 7577........	18 2 20	..	..	5'706	— 0'0010	..
800	..	.8	....	18 2 32'44	59'30	4	4'265	+ 0'0007	..
801	6156	6	Octantis.............	18 4 3'82	56'11	5	10'882	— 0'0203	..
802	6167	6	Telescopii............	18 5 20'27	57'56	1	5'058	— 0'0014	+ 0'014
803	6168	4	13 Sagittarii ...... μ	18 5 23'45	59'06	29	3'588	+ 0'0010	— 0'004
804	6179	5	15 Sagittarii..........	18 6 51	..	..	3'579	+ 0'0008	— 0'006
805	6180	6	16 Sagittarii.........	18 6 53	..	..	+ 3'570	+ 0'0008	+ 0'006

774. The R.A. has been brought up with Precession alone.
778. The R.A. has been brought up with Precession alone.
796. The R.A. has been brought up with Precession alone.

No.	Mean N.P.D. 1860, Jan. 1.	Mean Year and Fraction of Year.	No. of Obs. of N.P.D.	Annual Precess. in N.P.D. for 1860.	Secular Variation of Precess.in N.P.D.	Annual Proper Motion in N.P.D.	No. for reference.			
							Lacaille.	Brisbane.	Fallows or Johnson.	Greenwich or Henderson.
	° ′ ″	1800		″	″	″				
771	145 20 41·03	57·54	1	+ 2·01	− 0·726	+ 0·06	7413	6193	..	..
772	130 4 4·51	60·19	3	1·94	− 0·610	−. 0·03	7425	6198	284.J 447	H 69
773	117 46 23·35	59·02	9	1·86	− 0·549	+ 0·02	7440	..	J 448	1431
774	155 26 22·09	56·50	1	1·80	− 0·870	+ 0·05	7416	6201	..	..
775	121 39 2·42	59·65	1	1·74	− 0·567	+ 0·05	7451	..	..	..
776	126 59 36·80	59·58	1	1·72	− 0·593	− 0·05	7449	6214	J 449	..
777	127 28 52·83	59·32	5	1·72	− 0·596	..	..	..	..	..
778	130 2 24·96	56·50	1	1·71	− 0·611	+ 0·05	7447	6213	..	..
779	62 11 41·96	60·23	8	1·66	− 0·345	+ 0·74	..	..	..	1433
780	127 45 47·54	59·27	5	1·61	− 0·598	..	..	..	..	..
781	128 35 12·00	59·40	5	1·34	− 0·603	..	..	..	..	..
782	100 51 40·91	56·52	1	1·29	− 0·485	+ 0·18	..	..	..	1437
783	128 38 47·87	59·46	5	1·22	− 0·603	..	..	..	..	..
784	179 16 40·89	57·52	86	1·04	− 15·880	0·00	..	5912	275.J 423	H 3
785	120 14 3·07	59·53	11	0·87	− 0·561	+ 0·08	7521	..	J 450	1439
786	129 13 47·66	59·27	4	0·83	− 0·607	..	..	..	..	..
787	129 39 5·42	59·26	4	0·81	− 0·609	..	..	..	..	..
788	113 47 55·95	58·57	3	0·77	− 0·534	+ 0·04	7526	..	J 451	1440
789	38 29 26·04	56·49	1	0·58	− 0·203	+ 0·04	..	6294	286	1445
790	130 38 9·40	59·34	5	0·46	− 0·615	..	..	..	..	..
791	153 40 2·02	60·39	2	0·43	− 0·842	+ 0·19	7527	6291	..	..
792	130 26 52·04	59·32	4	0·41	− 0·614	..	..	..	..	..
793	98 10 34·59	56·52	1	0·40	− 0·476	− 0·02	..	..	J 455	1451
794	140 5 42·44	60·54	3	0·37	− 0·681	+ 0·04	7535	6296	287.J 454	..
795	119 34 54·93	58·48	4	0·34	− 0·559	+ 0·08	7552	..	J 456	1453
796	133 25 37·39	56·51	1	0·29	− 0·633	+ 0·13	7550	6302	..	..
797	120 25 16·54	59·48	5	0·28	− 0·563	+ 0·23	7557	..	J 457	1455
798	118 28 5·85	58·63	1	+ 0·07	− 0·554	+ 0·11	7579	..	J 458	1458
799	153 5 6·85	60·43	1	− 0·20	− 0·832	+ 0·05	7577	6329	..	..
800	131 44 28·41	59·30	4	0·22	− 0·622	..	..	..	..	..
801	170 17 13·29	56·11	5	0·36	− 1·587	− 0·01	7525	6324	..	..
802	146 3 39·91	57·56	1	0·47	− 0·738	+ 0·08	7608	6347	..	..
803	111 5 29·02	58·79	72	0·47	− 0·523	+ 0·01	..	..	288.J 460	1464
804	110 45 57·18	56·52	1	0·60	− 0·521	+ 0·02	..	..	..	1466
805	110 25 32·21	56·52	1	− 0·60	− 0·520	+ 0·03	..	..	..	2013*

No.	No. in B.A.C.	Magnitude	Star's Name.	Mean R.A. 1860, Jan. 1.	Mean Year and Fraction of Year	No. of Obs. R.A.	Annual Precess. in R.A. for 1860.	Secular Variation of Precess. in R.A.	Annual Proper Motion in R.A.
				h m s	1800		"	"	"
806	..	10·11	....	18 7 10·06	59·50	5	+4·297	—0·0005	..
807	..	8·9	....	18 7 14·21	59·50	5	4·286	—0·0005	..
808	6186	4	Sagittarii......... η	18 8 9·18	60·20	4	4·072	—0·0003	—0·016
809	..	6	Lacaille 7644.......	18 8 39·87	59·29	4	4·289	—0·0009	..
810	..	10	....	18 11 1·60	59·45	4	4·319	—0·0017	..
811	6209	3·4	19 Sagittarii...... δ	18 12 1·86	59·15	8	3·839	—0·0005	—0·001
812	..	8	....	18 12 17·84	59·29	4	4·317	—0·0020	..
813	6205	6	Lacaille 7562........	18 13 25·60	56·13	5	12·452	—0·0900	..
814	6228	6	Lacaille 7680.......	18 14 7·00	59·05	7	4·369	—0·0026	—0·007
815	6233	3·2	20 Sagittarii...... ε	18 14 52·70	57·51	19	3·987	—0·0015	—0·004
816	6240	4	Telescopii......... a	18 16 35·08	60·62	1	4·455	—0·0038	—0·014
817	6248	6	Telescopii.......	18 17 53·06	57·56	1	5·173	—0·0086	..
818	6250	4·5	Telescopii........ ζ	18 18 2·69	59·98	3	4·613	—0·0052	+0·015
819	..	9	....	18 18 21·40	59·30	6	4·370	—0·0039	..
820	..	6	Lacaille 7711........	18 19 3·67	59·28	4	4·356	—0·0041	..
821	6263	·3	22 Sagittarii ...... λ	18 19 19·83	59·96	10	3·707	—0·0012	·· 0·005
822	6275	7	Piazzi XVIII. 72...	18 20 49·83	59·62	2	3·941	—0·0024	..
823	6278	5	Telescopii......... δ¹	18 21 23·11	60·70	1	4·451	—0·0054	—0·002
824	..	8	....	18 21 48·08	59·38	5	4·388	—0·0050	..
825	6285	5·6	Lacaille 7746........	18 21 53·67	60·58	1	3·939	—0·0026	—0·002
826	6290	6	60 Serpentis........ c	18 22 24	..	..	3·120	+0·0005	0·000
827	6291	6	Telescopii............	18 23 15·40	57·70	1	4·836	—0·0087	..
828	6296	5	Coronæ Australis θ	18 23 30·05	60·41	1	4·287	—0·0049	—0·019
829	6307	6	61 Serpentis........ e	18 24 43	..	..	3·097	+0·0004	+0·002
830	6305	5·6	Lacaille 7761........	18 24 46·65	60·20	4	3·939	—0·0032	0·000
831	6312	6	24 Sagittarii.........	18 25 20·27	57·28	1	3·667	—0·0018	+0·002
832	6315	4	Pavonis............. ζ	18 26 39·72	60·57	2	7·050	—0·0386	—0·004
833	..	7	....	18 27 59·68	59·27	4	4·426	—0·0072	..
834	6328	6	Pavonis...............	18 28 41·28	57·56	1	5·886	—0·0231	—0·005
835	6343	6	Sagittarii...........	18 30 0	..	..	3·652	—0·0025	—0·002
836	6352	5	Lacaille 7785.......	18 31 41·56	60·32	5	5·911	—0·0263	—0·011
837	6355	1	3 Lyræ............. a	18 32 11·85	57·51	13	2·013	+0·0016	+0·017
838	..	10·11	....	18 32 44·57	60·60	2	3·877	—0·0042	..
839	..	10	....	18 33 20·67	59·36	4	4·457	—0·0092	..
840	6361	5	2 Aquilæ.............	18 34 36	..	..	+3·286	—·0010	+0·004

No.	Mean N P.D. 1860, Jan. 1.	Mean Year and Fraction of Year.	No. of Obs. of N.P.D.	Annual Precess. in N.P.D. for 1860.	Secular Variation of Precess. in N.P.D.	Annual Proper Motion in N.P.D.	No. for reference.			
							Lacaille.	Brisbane.	Fallows or Johnson.	Greenwich or Henderson.
	° ′ ″	1800		″	″	″				
806	132 30 47·97	59·51	4	− 0·62	− 0·627	··	··	··	··	··
807	132 15 27·86	59·51	4	0·63	− 0·625	··	··	··	··	··
808	126 47 56·62	60·20	4	0·71	− 0·594	+ 0·18	7643	6360	J 461	··
809	132 20 4·60	59·29	4	0·75	− 0·625	··	··	··	··	··
810	133 1 57·99	59·46	3	0·96	− 0·629	··	··	··	··	··
811	119 52 59·22	58·72	29	1·05	− 0·559	+ 0·04	7670	6377	J 462	1473
812	132 59 35·86	59·29	4	1·07	− 0·629	··	··	··	··	··
813	171 54 7·07	56·13	5	1·18	− 1·813	− 0·03	7562	6362	··	··
814	134 10 28·59	59·05	7	1·23	− 0·635	− 0·05	7680	6386	··	··
815	124 26 45·87	57·40	19	1·30	− 0·580	+ 0·14	7689	6391	290·J 464	1480
816	136 2 25·41	60·62	1	1·45	− 0·647	+ 0·07	7694	6397	J 465	··
817	147 36 11·37	57·56	1	1·56	− 0·751	+ 0·04	7696	6399	··	··
818	139 8 27·24	59·98	3	1·58	− 0·671	+ 0·23	7702	6403	··	··
819	134 14 40·52	59·30	6	1·60	− 0·634	··	··	··	··	··
820	133 55 43·94	59·28	4	1·66	− 0·632	··	··	··	··	··
821	115 29 41·63	59·39	20	1·69	− 0·538	+ 0·24	7725	··	J 467	1487
822	123 8 1·90	59·63	2	1·82	− 0·572	··	7735	··	··	1489
823	136 0 13·60	60·70	1	1·87	− 0·645	0·00	7729	6419	293·J 468	··
824	134 41 5·52	59·38	5	1·90	− 0·636	··	··	··	··	··
825	123 4 38·01	60·58	1	1·91	− 0·571	+ 0·12	7746	··	··	1493
826	92 4 20·11	56·51	2	1·96	− 0·451	+ 0·03	··	··	··	··
827	142 59 16·22	57·70	1	2·03	− 0·701	− 0·06	7743	6424	··	··
828	132 24 30·47	60·42	2	2·05	− 0·621	+ 0·02	7756	6427	J 471	··
829	91 5 56·18	58·59	2	2·16	− 0·448	+ 0·03	··	··	··	··
830	123 6 56·76	60·20	4	2·17	− 0·571	··	7761	··	··	··
831	114 7 56·45	57·28	1	2·21	− 0·531	0·00	7769	··	··	1497
832	161 32 23·72	60·57	2	2·33	− 1·021	+ 0·13	7736	6436	295·J 472	··
833	135 34 39 42	59·27	4	2·44	− 0·640	··	··	··	··	··
834	154 45 45·70	57·56	1	2·50	− 0·851	+ 0·08	7766	6446	··	··
835	113 37 13·07	56·52	1	2·62	− 0·527	+ 0·03	7806	··	··	··
836	154 59 44·84	60·32	5	2·77	− 0·853	+ 0·10	7785	6458	J 473	··
837	51 20 40·95	57·74	16	2·81	− 0·290	− 0·28	··	6466	··	1501
838	121 17 42·54	60·60	2	2·86	− 0·559	··	··	··	··	··
839	136 18 19·11	59·36	4	2·90	− 0·642	··	··	··	··	··
840	99 10 57·81	60·66	3	− 3·02	− 0·472	0·00	··	··	297·J 474	1502

No.	No. in B.A.C.	Magnitude.	Star's Name.	Mean R.A. 1860, Jan. 1.	Mean Year and Fraction of Year.	No. of Obs. R.A.	Annual Precess. in R.A. for 1860.	Secular Variation of Precess. in R.A.	Annual Proper Motion in R.A.
				h m s	1800		s	s	s
841	6360	5	Pavonis .... θ	18 34 51·14	60·39	6	+ 5·935	− 0·0294	− 0·007
842	··	9	....	18 35 54·17	59·58	5	4·475	− 0·0102	··
843	6371	4·3	27 Sagittarii ...... φ	18 36 54·52	59·24	15	3·748	− 0·0040	+ 0·004
844	6379	5·6	4 Aquilæ	18 37 46	··	··	3·028	0·0000	+ 0·005
845	6383	5	Pavonis .... λ	18 39 13·88	60·39	3	5·586	− 0·0269	− 0·008
846	6385	6	Coronæ Australis η²	18 39 30·38	57·70	1	4·326	− 0·0095	− 0·007
847	··	6	Lacaille 7872	18 42 2·37	59·30	5	4·470	− 0·0121	··
848	6405	5	Pavonis .... κ	18 42 29·59	60·27	3	6·230	− 0·0423	− 0·011
849	··	6·7	Lacaille 7889	18 43 32·86	59·31	4	4·502	− 0·0130	··
850	6414	6·7	Lacaille 7898	18 43 42·35	60·66	2	3·858	− 0·0060	··
851	··	9	....	18 44 37·67	59·34	4	4·518	− 0·0136	··
852	6429	Var.	10 Lyræ .... β¹	18 44 54·73	57·06	10	2·214	+ 0·0015	− 0·002
853	6440	2·3	34 Sagittarii .... σ	18 46 34·95	59·55	26	3·724	− 0·0052	0·000
854	··	7·8	....	18 46 35·96	59·34	5	4·514	− 0·0141	··
855	6442	6	Coronæ Australis...	18 47 8·01	57·70	1	4·340	− 0·0119	··
856	6443	6	Telescopii .... λ	18 47 15·32	40·61	1	4·816	− 0·0184	··
857	··	9	....	18 48 22·67	59·44	5	4·532	− 0·0150	··
858	6451	6	62 Serpentis	18 48 37	··	··	2·924	− 0·0001	+ 0·005
859	6461	4	37 Sagittarii ...... ξ²	18 49 22·58	60·50	1	3·581	− 0·0044	− 0·001
860	··	7	Lacaille 7930	18 50 3·74	59·27	4	4·546	− 0·0157	··
861	6489	3·4	38 Sagittarii .... ζ	18 53 42·11	58·75	23	3·825	− 0·0074	− 0·005
862	··	7	Lacaille 7959	18 54 4·77	59·36	5	4·547	− 0·0172	··
863	··	9	....	18 54 19·23	59·41	5	4·559	− 0·0174	··
864	6506	6	Sagittarii	18 56 21·94	57·70	1	4·538	− 0·0176	··
865	··	11	....	18 56 50·60	59·48	4	4·576	− 0·0187	··
866	6511	5	Coronæ Australis γ	18 56 56·96	60·38	2	4·058	− 0·0108	+ 0·004
867	6521	4·3	40 Sagittarii .... τ	18 58 11·80	59·33	16	3·756	− 0·0073	− 0·008
868	6523	5	Coronæ Australis δ	18 58 35·89	59·63	1	4·185	− 0·0130	+ 0·002
869	6528	3	17 Aquilæ .... ζ	18 58 58·55	59·00	19	2·758	+ 0·0003	− 0·006
870	··	8·9	....	18 59 31·62	59·40	4	4·587	− 0·0198	··
871	6535	4·5	Coronæ Australis α	18 59 56·65	60·24	3	4·085	− 0·0118	+ 0·007
872	6541	5	Coronæ Australis β	19 0 23·61	60·22	3	4·138	− 0·0128	− 0·005
873	6548	4·5	41 Sagittarii ...... π	19 1 26·10	60·72	1	3·573	− 0·0057	− 0·004
874	··	10	....	19 4 1·56	59·31	5	4·591	− 0·0215	··
875	6575	6	42 Sagittarii ...... ψ	19 6 57·22	60·46	13	+ 3·683	− 0·0077	0·000

856. The R.A. has been brought up with Precession alone.

No.	Mean N.P.D. 1860, Jan. 1.	Mean Year and Fraction of Year.	No. of Obs. of N.P.D.	Annual Precess. in N.P.D. for 1860.	Secular Variation of Precess.in N.P.D.	Annual Proper Motion in N.P.D.	No. for reference.			
							Lacaille.	Brisbane.	Fallows or Johnson.	Greenwich or Henderson.
	° ′ ″	1800		″	″	″				
841	155 12 56·78	60·39	6	− 3·04	− 0·855	+ 0·04	7813	6467	..	..
842	136 43 36·40	59·38	5	3·12	− 0·644	..	..	..	..	..
843	117 7 49·52	59·05	24	3·22	− 0·538	− 0·01	7844	6482	298.J 475	1503
844	88 4 44·60	56·51	1	3·29	− 0·434	− 0·02	..	..	..	2054*
845	152 20 29·37	60·20	4	3·42	− 0·801	+ 0·04	7841	6489	J 476	..
846	133 35 0·65	57·70	1	3·41	− 0·620	− 0·03	7859	6493	..	..
847	136 45 15·66	59·30	5	3·65	− 0·639	..	..	···	..	..
848	157 24 7·03	60·27	3	3·70	− 0·891	− 0·10	7856	6503	..	..
849	137 26 15·54	59·31	4	3·78	− 0·643	..	7889	..	..	..
850	120 53 44·97	60 66	2	3·80	− 0·551	..	7898	..	..	1515
851	137 47 9·72	59·34	4	3·88	− 0·644	..	..	..	..	..
852	56 47 51·01	57·03	11	3·91	− 0·315	+ 0·03	..	..	..	1518
853	116 27 59·31	59·26	40	4·05	− 0·530	+ 0·08	7918	6527	300.J 478	1521
854	137 45 10·85	59·34	5	4·04	− 0·643	..	..	..	..	..
855	134 5 30·62	57·70	1	4·10	− 0·618	+ 0·01	7914	6530	..	..
856	143 7 0·89	56·51	1	4·11	− 0·686	0·00	7910	6528	..	..
857	138 9 15·66	59·44	5	4·20	− 0·644	..	..	..	..	..
858	83 33 24·22	56·51	1	4·22	− 0·415	+ 0·06	..	..	..	2080*
859	111 17 12·10	60·50	1	4·29	− 0·508	+ 0·03	..	..	J 480	1533
860	138 28 12·90	59·28	4	4·34	− 0·645	..	..	..	..	..
861	120 4 33·71	58·75	23	4·66	− 0·541	+ 0·03	7966	..	301.J 481	1540
862	138 36 8·78	59·36	5	4·68	− 0·643	..	..	..	..	..
863	138 52 4·31	59·41	5	4·70	− 0·644	..	..	..	..	..
864	138 30 22·49	57·70	1	4·88	− 0·641	+ 0·05	7973	6569	..	..
865	139 14 11·59	59·48	4	4·92	− 0·645	..	..	..	..	..
866	127 15 35·71	60·38	2	4·93	− 0·572	+ 0·29	7988	6574	J 483	..
867	117 52 15·02	59·32	18	5·04	− 0·528	+ 0·26	7994	..	302.J 484	1544
868	130 42 32·08	59·63	1	5·07	− 0·589	+ 0·07	7992	6578	J 485	..
869	76 20 30·32	59·08	37	5 10	− 0·387	+ 0·07	..	..	..	1545
870	139 31 50·35	59·40	4	5·15	− 0·645	..	..	..	..	..
871	128 7 3·99	60·24	3	5·19	− 0·574	+ 0·11	8002	6585	J 487	..
872	129 33 29·82	60·22	3	5·22	− 0·581	+ 0·07	8007	6587	J 488	..
873	111 14 32·46	60·73	2	5·31	− 0·500	+ 0·03	..	6594	304.J 489	1548
874	139 46 13·32	59·31	5	5·53	− 0·642	..	..	..	..	..
875	115 29 38·67	60·48	14	− 5·78	− 0·512	+ 0·01	8052	..	..	1550

No.	No. in B.A.C.	Magnitude	Star's Name.	Mean R.A. 1860, Jan. 1.	Mean Year and Fraction of Year.	No. of Obs. of R.A.	Annual Precess. in R.A. for 1860.	Secular Variation of Precess. in R.A.	Annual Proper Motion in R.A.
				h m s	1800		s	s	s
876	6595	6·5	25 Aquilæ......... ω	19 11 14·69	58·34	16	+ 2·817	− 0·0003	− 0·003
877	..	9	....	19 12 20·91	59·33	4	4·614	− 0·0252	..
878	6608	3·4	Sagittarii......... β¹	19 12 33·91	60·18	5	4·330	− 0·0192	− 0·003
879	6610	4	Sagittarii......... β²	19 13 5·68	60·52	2	4·344	− 0·0197	+ 0·004
880	6619	4	44 Sagittarii...... ρ¹	19 13 32·85	60·65	1	3·487	− 0·0061	− 0·003
881	6622	4	Sagittarii.......... a	19 14 10·65	60·02	2	4·169	− 0·0167	− 0·011
882	..	11	....	19 14 42·87	59·40	4	4·624	− 0·0263	..
883	6636	6	49 Sagittarii...... χ²	19 17 1·09	58·69	2	3·640	− 0·0085	+ 0·002
884	6639	6	Lacaille 8107.......	19 18 5·41	60·26	6	3·800	− 0·0110	+ 0·004
885	6646	3·4	30 Aquilæ......... δ	19 18 26·34	60·02	21	3·010	− 0·0017	+ 0·014
886	6649	4	Telescopii......... μ	19 19 12·12	60·03	2	4·895	− 0·0352	− 0·009
887	..	10	....	19 19 20·27	59·33	4	4·628	− 0·0282	..
888	6666	6	Piazzi XIX. 126...	19 21 12·26	57·74	3	3·718	− 0·0102	− 0·006
889	..	..	....	19 22 1·97	60·72	1	3·785	− 0·0115	..
890	..	..	....	19 22 17·43	60·56	1	3·788	− 0·0115	..
891	..	11	....	19 23 43·93	59·40	5	4·646	− 0·0303	..
892	6682	7	Lacaille 8139........	19 23 56·55	60·61	11	3·744	− 0·0110	..
893	..	10	....	19 27 2·62	59·56	6	4·647	− 0·0317	..
894	6706	5·4	52 Sagittarii...... h²	19 28 11·04	59·95	25	3·655	− 0·0101	+ 0·002
895	..	7	....	19 29 53·29	59·44	5	4·646	− 0·0328	..
896	6708	6	Octantis.............	19 29 59·84	56·11	5	11·528	− 0·5149	..
897	..	9	....	19 30 44·94	59·50	5	4·656	− 0·0335	..
898	..	10·11	....	19 33 10·09	59·46	5	4·652	− 0·0343	..
899	..	11	....	19 33 25·35	59·47	4	4·659	− 0·0346	..
900	6753	6·7	Lacaille 8208........	19 36 32·06	60·01	9	3·812	− 0·0142	− 0·005
901	..	9	....	19 37 41·32	60·70	4	3·708	− 0·0124	..
902	6760	5	56 Sagittarii........ f	19 38 11·57	59·85	7	3·517	− 0·0091	− 0·013
903	..	10·11	....	19 38 20·68	59·35	4	4·652	− 0·0364	..
904	6770	7	Lacaille 8225........	19 39 5·86	60·62	10	3·740	− 0·0133	..
905	6772	3	50 Aquilæ......... γ	19 39 36·15	59·46	9	2·852	− 0·0010	+ 0·001
906	6786	7	Lacaille 8241........	19 41 46·76	60·54	1	3·689	− 0·0126	..
907	..	10·11	....	19 42 11·30	59·38	5	4·654	− 0·0382	..
908	6792	7	Lacaille 8243........	19 42 32·27	60·58	3	3·708	− 0·0131	..
909	6802	1·2	53 Aquilæ......... a	19 43 57·10	59·16	42	2·892	− 0·0014	+ 0·036
910	6801	4	Pavonis............. ε	19 44 20·15	57·13	7	+ 7·068	− 0·1623	+ 0·036

No.	Mean N.P.D. 1860, Jan. 1.	Mean Year and Fraction of Year.	No. of Obs. of N.P.D.	Annual Precess. in N.P.D. for 1860.	Secular Variation of Precess.in N.P.D.	Annual Proper Motion in N.P.D.	No. for reference.			
							Lacaille.	Brisbane.	Fallows or Johnson.	Greenwich or Henderson.
876	78 39 15·12	1800 58·69	24	− 6·14	− 0·389	− 0·02	..	..	..	1558
877	140 30 8·06	59·33	4	6·22	− 0·637	..	..	..	..	..
878	134 43 1·59	60·18	5	6·24	− 0·597	+ 0·02	8075	..	306.J 492	..
879	135 3 29·89	60·52	2	6·29	− 0·598	+ 0·07	8079	..	..	..
880	108 6 25·04	60·65	1	6·33	− 0·480	− 0·03	..	..	308.J 493	1563
881	130 52 29·25	60·02	2	6·38	− 0·573	+ 0·07	8087	6650	310.J 494	..
882	140 46 45·03	59·41	5	6·42	− 0·636	..	..	..	..	..
883	114 13 58·77	58·69	2	6·61	− 0·498	− 0·04	8103	..	..	1567
884	120 0 57·91	60·26	6	6·70	− 0·519	+ 0·08	8107	..	..	..
885	87 9 40·74	59·91	36	6·73	− 0·411	− 0·10	..	..	311	1570
886	145 23 30·07	60·24	3	6·80	− 0·669	0·00	8101	6666	..	..
887	141 2 51·86	59·33	4	6·80	− 0·632	..	..	..	..	..
888	117 16 5·66	57·74	3	6·96	− 0·506	+ 0·13	8123	..	..	1573
889	119 40 7·84	60·72	1	7·03	− 0·514	..	..	..	..	..
890	119 46 51·64	60·56	1	7·05	− 0·514	..	..	..	..	..
891	141 34 32·69	59·41	4	7·16	− 0·630	..	..	..	..	..
892	118 17 2·71	60·61	11	7·18	− 0·507	..	8139	..	..	1575
893	141 44 52·69	59·56	6	7·43	− 0 627	..	..	..	..	..
894	115 11 19·70	59·45	46	7·53	− 0·491	− 0·02	8166	..	J 497	1582
895	141 51 45·72	59·46	6	7·66	− 0·623	..	..	..	..	..
896	171 41 20·49	56·11	5	7·67	− 1·552	− 0·04	8094	6694	..	..
897	142 5 28·07	59·50	5	7·73	− 0·624	..	..	..	..	..
898	142 7 52·92	59·46	5	7·93	− 0·620	..	..	..	..	..
899	142 16 8·38	59·47	4	7·95	− 0·621	..	..	..	..	..
900	121 14 5·91	60·01	9	8·20	− 0·504	+ 0·12	8208	..	..	1594
901	117 36 10·80	60·70	4	8·30	− 0·488	..	..	..	..	..
902	110 5 38·87	59·85	7	8·33	− 0·463	+ 0·07	..	6734	..	1598
903	142 25 4·84	59·35	4	8·34	− 0·613	..	..	..	..	..
904	118 49 52·93	60·62	9	8·41	− 0·491	..	8225	..	..	..
905	79 43 30·23	59·78	12	8·44	− 0·374	0·00	..	6742	318	1600
906	117 3 49·66	60·54	2	8·62	− 0·482	..	8241	..	..	1605
907	142 40 2·30	59·38	5	8·64	− 0·609	..	..	..	..	..
908	117 49 19·71	60·58	3	8·68	− 0·484	..	8243	..	..	1607
909	81 29 54·71	58·74	108	8·79	− 0·376	− 0·38	..	6758	320	1610
910	163 16 22·34	57·02	7	− 8·82	− 0·922	+ 0·13	8219	6752	J 501	..

No.	No. in B.A.C.	Magnitude	Star's Name.	Mean R.A. 1860, Jan. 1.	Mean Year and Fraction of Year.	No. of Obs. R.A.	Annual Precess. in R.A. for 1860.	Secular Variation of Precess. in R.A.	Annual Proper Motion in R.A.
				h m s	1800		s	s	s
911	6812	4·5	Sagittarii............ ι	19 45 35·60	59·56	6	+ 4·159	− 0·0242	− 0·003
912	..	10	....	19 45 46·04	59·53	5	4·665	− 0·0400	..
913	6823	5	58 Sagittarii...... ω	19 47 15·54	60·67	5	3·671	− 0·0130	+ 0·013
914	6832	5	59 Sagittarii....... b	19 48 21·11	58·93	13	3·693	− 0·0136	− 0·003
915	6833	4	60 Aquilæ......... β	19 48 26·17	59·69	9	2·946	− 0·0020	+ 0·002
916	6842	5	60 Sagittarii...... A	19 50 25·17	60·04	6	3·665	− 0·0133	− 0·004
917	..	9	....	19 50 42·62	59·37	5	4·664	− 0·0421	..
918	..	8·9	....	19 50 53·15	59·35	5	4·655	− 0·0418	..
919	..	9	....	19 52 9·54	58·38	4	3·531	− 0·0108	..
920	..	8	....	19 53 4·57	56·18	1	9·739	− 0·4341	..
921	6870	5	62 Sagittarii c	19 54 2·66	59·99	10	3·699	− 0·0146	0·000
922	..	..	................:	19 54 30·32	60·63	7	3·655	− 0·0137	..
923	6859	6	Octantis....:.......	19 54 31·71	56·14	5	13·751	− 1·0427	..
924	6873	4	Pavonis........... δ	19 54 57·13	59·63	5	5·772	− 0·0958	+ 0·190
925	6877	5	Lacaille 8322.......	19 55 26·74	60·44	2	3·816	− 0·0175	+ 0·023
926	..	7	Lacaille 8334.......	19 56 39·25	60·63	5	3·672	− 0·0143	..
927	6900	6	Octantis.............	19 59 56·37	56·20	5	9·652	− 0·4545	..
928	..	10·11	....	20 0 21·98	58·38	4	3·522	− 0·0115	..
929	..	9	....	20 0 24·87	59·38	4	4·650	− 0·0457	..
930	6906	7	Lacaille 8359........	20 0 34·99	60·62	4	3·652	− 0·0144	..
931	6920	7	Lacaille 8364.........	20 1 40·17	60·69	4	3·627	− 0·0139	..
932	6922	6	Lacaille 8362..	20 1 59·63	60·05	4	3·923	− 0·0216	..
933	..	10	....	20 2 42·98	59·33	5	4·656	− 0·0471	..
934	..	10	....	20 4 5·44	58·35	3	3·515	− 0·0117	..
935	..	10	....	20 5 27·05	59·44	5	4·654	− 0·0483	..
936	..	9·10	....	20 6 58·11	59·51	4	4·651	− 0·0488	..
937	6948	7	Lacaille 8386........	20 7 8·46	60·55	2	3·740	− 0·0175	..
938	..	11	....	20 9 9·72	58·39	3	3·512	− 0·0122	..
939	..	12	....	20 9 25·29	59·35	4	4·655	− 0·0497	..
940	..	..	....	20 9 44·59	60·63	6	3·593	− 0·0141	..
941	6971	6	4 Capricorni.........	20 9 47·69	59·88	3	3·533	− 0·0127	+ 0·006
942	6974	3·4	6 Capricorni...... a²	20 10 17·05	59·34	26	3·332	− 0·0084	+ 0·001
943	6981	6·5	7 Capricorni ...... σ	20 11 18·75	59·48	3	3·471	− 0·0114	0·000
944	6982	7	Lacaille 8407........	20 11 22·27	60·52	2	3·611	− 0·0148	..
945	..	..	....	20 12 22·63	60·65	5	+ 3·600	− 0·0146	..

No.	Mean N.P.D. 1860, Jan. 1.	Mean Year and Fraction of Year.	No. of Obs. of N.P.D.	Annual Precess. in N.P.D. for 1860.	Secular Variation of Precess. in N.P.D.	Annual Proper Motion in N.P.D.	No. for reference.			
							Lacaille.	Brisbane.	Fallows or Johnson.	Greenwich or Henderson
	° ′ ″	1800		″	″	″				
911	132 13 54·16	59·57	7	− 8·92	− 0·539	− 0·08	8255	..	J 502	..
912	143 4 35·90	59·53	5	8·93	− 0·605	..	..	..	..	..
913	116 40 2·08	60·67	5	9·05	− 0·474	− 0·08	8268	..	..	1616
914	117 32 14·17	58·93	13	9·13	− 0·476	+ 0·01	8277	..	321.J 503	1619
915	83 56 24·19	59·95	8	9·14	− 0·379	+ 0·47	..	6774	322	1620
916	116 34 15·09	60·04	6	9·29	− 0·470	− 0·01	8294	..	..	1623
917	143 21 32·48	59·37	5	9·31	− 0·599	..	..	..	..	..
918	143 12 21·25	59·35	2	9·32	− 0·597	..	..	..	..	..
919	111 14 7·38	58·37	5	9·43	− 0·451	..	..	..	..	..
920	169 58 58·83	56·18	1	9·50	− 1·251	..	..	..	..	..
921	118 5 43·78	58·56	21	9·57	− 0·470	− 0·02	8315	..	323.J 505	1627
922	116 25 40·30	60·63	7	9·61	− 0·464	..	..	..	..	..
923	173 43 46·36	56·14	5	9·61	− 1·757	+ 0·03	8202	6771	..	..
924	156 31 57·56	59·63	5	9·64	− 0·735	+ 1·15	8295	6787	J 504	..
925	122 26 43·01	60·44	2	9·68	− 0·483	+ 0·09	8322	..	J 506	1629
926	117 12 20·11	60·63	5	9·77	− 0·463	..	8334	..	..	1631
927	170 1 7·34	56·20	5	10·02	− 1·216	− 0·02	8281	6796	..	..
928	111 14 1·97	58·37	4	10·06	− 0·441	..	..	..	..	..
929	143 44 43·50	59·38	4	10·06	− 0·583	..	..	..	..	..
930	116 37 31·30	60·62	4	10·07	− 0·456	..	8359	..	..	1636
931	115 41 28·14	60·69	4	10·15	− 0·452	0·00	8364	..	..	1637
932	126 26 56·85	60·05	4	10·18	− 0·489	..	8362	..	..	..
933	144 1 11·35	59·33	5	10·23	− 0·580	..	..	..	..	..
934	111 8 33·62	58·35	3	10·34	− 0·435	..	..	..	..	..
935	144 10 39·87	59·44	5	10·43	− 0·575	..	.	..	..	..
936	144 14 32·73	59·51	4	10·55	− 0·572	..	..	..	..	..
937	120 25 44·56	60·55	2	10·57	− 0·459	..	8386	..	..	1645
938	111 15 43·66	58·39	3	10·72	− 0·428	..	..	..	..	..
939	144 29 28·08	59·35	4	10·73	− 0·569	..	..	..	..	..
940	114 48 0·88	60·63	6	10·76	− 0·437	..	..	..	..	..
941	112. 14 20·30	59·88	3	10·76	− 0·430	+ 0·05	..	..	..	1656
942	102 58 32·43	59·46	86	10·80	− 0·405	0·00	..	..	326.J 509	1660
943	109 33 8·28	59·48	3	10·87	− 0·421	− 0·02	..	..	327	1663
944	115 39 28·42	60·52	2	10·88	− 0·438	..	8407	..	..	1664
945	115 14 8·20	60·65	5	− 10·96	− 0·435	..	..	..	..	..

No.	No. in B.A.C.	Magnitude.	Star's Name.	Mean R.A. 1860, Jan. 1.	Mean Year and Fraction of Year.	No. of Obs. R.A.	Annual Precess. in R.A. for 1860.	Secular Variation of Precess. in R.A.	Annual Proper Motion in R.A.
				h m s	1800		s	s	s
946	6995	3	9 Capricorni......β	20 13 8·39	59·83	1	+3·376	-0·0095	-0·001
947	..	9	....	20 13 22·62	58·34	3	3·505	-0·0125	..
948	7004	2	Pavonis.........a	20 14 32·84	59·59	13	4·798	-0·0593	-0·003
949	..	..	....	20 14 40·36	60·63	5	3·601	-0·0149	..
950	7011	7	Lacaille 8427........	20 16 6·38	60·46	3	3·700	-0·0177	..
951	7012	7	Lacaille 8430........	20 16 11·57	60·52	2	3·619	-0·0156	..
952	..	10·11	....	20 16 35·08	59·36	4	4·632	-0·0524	..
953	7022	3	37 Cygni..........γ	20 17 12·29	60·68	2	2·151	+0·0019	0·000
954	7021	7	Lacaille 8440........	70 17 21·24	60·53	1	3·635	-0·0162	..
955	..	9	....	20 17 22·77	59·48	5	4·635	-0·0530	..
956	..	11	....	20 19 16·36	59·35	4	4·462	-0·0540	..
957	7039	7	Lacaille 8458......	20 20 9·93	60·48	6	3·574	-0·0149	..
958	7040	7	Lacaille 8459........	20 20 26·09	60·46	2	3·569	-0·0148	..
959	7042	5	11 Capricorni.....ρ	20 20 52·30	60·10	20	3·433	-0·0114	-0·006
960	..	8	....	20 22 6·45	59·35	5	4·625	-0 0548	..
961	7057	6·7	Lacaille 8466........	20 22 21·93	60·53	1	3·689	-0·0184	..
962	7062	5·6	43 Cygni.........ω¹	20 22 45·38	60·68	2	1·826	+0·0001	+0·005
963	..	10·11	....	20 23 8·26	59·40	4	4·617	-0·0549	..
964	7077	6	Lacaille 8480........	20 24 31·86	59·63	7	3·585	-0·0157	-0·011
965	7068	5·6	Octantis..........μ¹	20 24 37·42	56·31	5	7·626	-0·3009	+0·072
966	..	10	....	20 25 31·04	59·45	4	4·613	-0·0558	..
967	7091	6·5	46 Cygni.........ω³	20 26 59·44	60·68	2	1·850	+0·0004	+0·002
968	7096	3	Indi...............a	20 27 42·21	60·12	9	4·250	-0·0398	0·000
969	7106	5	Pavonis............ν	20 29 3·36	60·03	4	5·610	-0·1199	+0·002
970	7119	6	Cygni...............	20 30 18·49	60·70	1	2·138	+0·0023	..
971	..	9·10	....	20 30 32·49	58·35	4	3·475	-0·0133	..
972	7127	5	14 Capricorni.... τ²	20 31 26·41	60·10	4	3·364	-0·0106	-0·002
973	..	11	....	20 31 28·52	59·41	5	4·614	-0·0587	..
974	7131	6·5	15 Capricorni.....υ	20 32 4·62	58·60	4	3·427	-0·0122	0·000
975	7129	3	Pavonis............β	20 32 17·63	59·70	2	5·516	-0·1162	-0·009
976	..	11·12	....	20 33 45·60	59·41	5	4·602	-0·0591	..
977	..	11·12	....	20 34 39·65	59·48	5	4·602	-0·0598	..
978	..	6·5	Lacaille 8537........	20 34 44·27	60·47	4	3·672	-0·0196	..
979	..	8·9	....	20 35 1·20	58·34	3	3·467	-0·0136	..
980	7165	4·5	Pavonis............σ	20 35 59·31	57·91	9	+5·819	-0·1443	..

No.	Mean N P.D. 1860, Jan. 1.	Mean Year and Fraction of Year.	No. of Obs. of N.P.D.	Annual Precess. in N.P.D. for 1860.	Secular Variation of Precess. in N.P.D.	Annual Proper Motion in N.P.D.	No. for reference.			
							Lacaille.	Brisbane.	Fallows or Johnson.	Greenwich or Henderson.
	° ′ ″	1800		″	″	″				
946	105 13 13·45	60·43	3	− 11·01	− 0·407	− 0·03	..	..	328.J 511	1667
947	111 10 30·07	58·34	3	11·03	− 0·422	..	..	..	..	..
948	147 10 45·22	59·50	16	11·11	− 0·578	+ 0·08	8416	6846	329.J 512	H 38
949	115 25 55·55	60·63	5	11·12	− 0·432	..	..	..	..	..
950	119 31 26·43	60·46	3	11·22	− 0·442	..	8427	..	..	1671
951	116 16 52·98	60·52	2	11·23	− 0·433	..	8430	..	..	1672
952	144 38 39·98	59·36	4	11·25	− 0·554	..	..	..	..	..
953	50 11 22	..	..	11·31	− 0·254	− 0·02	..	..	..	R 1972
954	117 0 28·69	60·53	1	11·31	− 0·433	..	8440	..	..	..
955	144 45 23·70	59·48	5	11·31	− 0·553	..	..	..	..	..
956	145 1 40·50	59·35	4	11·45	− 0·551	..	..	..	..	..
957	114 37 6·04	60·48	6	11·52	− 0·422	..	8458	..	..	1678
958	114 26 30·68	60·46	2	11·54	− 0·421	..	8459	..	..	1679
959	108 16 24·51	59·43	46	11·57	− 0·404	+ 0·01	..	..	331.J 514	1680
960	144 59 3·13	59·35	5	11·65	− 0·544	..	..	..	..	..
961	119 34 40·58	60·53	1	11·67	− 0·433	..	8466	..	..	1683
962	41 4 46	..	..	11·70	− 0·211	− 0·04	..	..	..	1684
963	144 55 38·99	59·40	4	11·72	− 0·541	..	..	..	..	..
964	115 24 47·57	59·63	7	11·83	− 0·417	+ 0·12	8480	..	..	1687
965	166 39 50·44	56·31	5	11·83	− 0·893	+ 0·14	8435	6870	..	..
966	145 3 9·26	59·45	4	11·89	− 0·537	..	..	..	..	..
967	41 15 3	..	..	12·00	− 0·210	+ 0·03	..	..	..	1691
968	137 46 32·54	60·18	11	12·05	− 0·491	− 0·09	8494	6885	335.J 516	H 52
969	157 14 57·00	60·03	4	12·14	− 0·646	+ 0·03	8488	6889	J 517	..
970	48 35 35	..	..	12·23	− 0·242	..	..	..	..	..
971	110 49 0·94	58·34	3	12·25	− 0·396	..	..	..	..	..
972	105 26 34·85	60·10	4	12·31	− 0·382	+ 0·03	..	..	..	1700
973	145 36 12·08	59·41	4	12·31	− 0·526	..	..	..	..	..
974	108 37 42·44	58·60	4	12·35	− 0·388	− 0·02	..	..	J 520	1701
975	156 42 4·40	60·04	3	12·37	− 0·628	+ 0·06	8500	6897	338.J 518	H 19
976	145 35 49·97	59·41	5	12·46	− 0·521	..	..	..	..	..
977	145 41 35·33	59·48	5	12·53	− 0·519	..	..	..	..	..
978	119 54 51·89	60·47	4	12·54	− 0·413	..	8537	..	..	..
979	110 43 23·07	58·34	3	12·56	− 0·389	..	..	..	..	..
980	159 16 56·49	57·65	4	− 12·62	− 0·655	− 0·03	8521	6908	..	..

No.	No. in B.A.C.	Magnitude.	Star's Name.	Mean R.A. 1860, Jan. 1.	Mean Year and Fraction of Year.	No. of Obs. of R.A.	Annual Precess. in R.A. for 1860.	Secular Variation of Precess. in R.A.	Annual Proper Motion in R.A.
				b m s	1800		s	s	s
981	7171	2·1	50 Cygni.......... a	20 36 39·44	59·48	6	+ 2·043	+ 0·0022	— 0·002
982	7020	6·7	Octantis.......... B	20 37 20·79	59·36	50	117·266	—157·5812	— 0·146
983	7177	4·5	16 Capricorni.... ψ	20 37 48·03	58·44	9	3·570	— 0·0168	— 0·007
984	7207	4·5	Microscopii....... a	20 41 12·98	59·90	4	3·767	— 0·0240	+ 0·021
985	7208	5·6	Indi............... ι	20 41 21·54	60·15	3	4·382	— 0·0511	— 0·001
986	7227	4·5	18 Capricorni..... ω	20 43 27·58	56·78	5	3·597	— 0·0184	— 0·003
987	7228	4	Indi.............. β	20 43 50·14	60·21	6	4·751	— 0·0734	— 0·008
988	7233	5·6	55 Cygni............	20 44 10·00	60·68	2	2·042	+ 0·0022	..
989	..	9	....	20 44 34·54	59·43	5	4·564	— 0·0626	..
990	7249	6	19 Capricorni........	20 46 52·96	59·38	2	3·405	— 0·0128	0·000
991	..	10·11	....	20 47 5·84	59·52	3	4·564	— 0·0638	..
992	..	9	....	20 48 0·08	59·37	4	4·564	— 0·0643	..
993	7254	6	Cygni...............	20 48 25·12	60·68	2	2·092	+ 0·0029	..
994	7256	5·6	32 Vulpeculæ........	20 48 35·67	58·90	4	2·555	+ 0·0026	— 0·002
995	..	11·12	....	20 48 49·15	58·56	4	3·441	— 0·0140	..
996	..	9·10	....	20 49 2·51	58·37	4	3·436	— 0·0138	..
997	7270	6	20 Capricorni.......	20 51 38·44	58·34	2	3·420	— 0·0136	+ 0·006
998	7277	4	58 Cygni.......... ν	20 51 57·09	60·68	2	2·233	+ 0·0037	+ 0·001
999	7282	6	21 Capricorni........	20 52 58·84	58·56	2	3·390	— 0·0128	0·000
1000	7292	5·6	Microscopii........ ζ	20 54 0·50	60·27	8	3·862	— 0·0303	·- 0·008
1001	7301	5·6	59 Cygni .... f'	20 55 3·84	60·68	2	2·037	+ 0·0031	— 0·005
1002	7305	5·6	22 Capricorni.... η	20 56 25·89	58·09	3	3·429	— 0·0143	— 0·006
1003	7309	5·6	12 Aquarii..........	20 56 40	..	..	3·179	— 0·0071	+ 0·006
1004	7314	5·6	Microscopii........ η	20 57 18·04	60·21	3	3·931	— 0·0342	— 0·002
1005	..	8·9	....	20 57 50·43	58·35	3	3·420	— 0·0141	..
1006	7322	4	23 Capricorni..... θ	20 58 4·35	59·71	6	3·378	— 0·0128	+ 0·004
1007	7328	5	24 Capricorni..... A	20 58 (55·94)	..	..	3·526	— 0·0178	+ 0·002
1008	7333	4	62 Cygni.......... ξ	20 59 50·27	60·68	2	2·178	+ 0·0041	+ 0·001
1009	7331	5·6	Pavonis............ o	21 0 8·48	56·17	5	5·769	— 0·1721	+ 0·010
1010	7335	6	25 Capricorni..... χ	21 0 32·18	57·59	2	3·448	— 0·0153	— 0·001
1011	7336	5·6	61 Cygni, 1st Star..	21 0 37·33	60·05	2	2·334	+ 0·0043	+ 0·339
1012	..	10	....	21 1 22·24	58·40	3	3·412	— 0·0141	..
1013	..	7	Lacaille 8707........	21 1 41·15	60·46	2	3·620	— 0·0217	..
1014	7344	4·5	13 Aquarii........ ν	21 1 57·86	60·43	5	3·270	— 0·0098	+ 0·001
1015	..	9	....	21 2 19·09	59·42	4	+ 4·518	— 0·0692	..

No.	Mean N.P.D. 1860, Jan. 1.	Mean Year and Fraction of Year.	No. of Obs. of N.P.D.	Annual Precess. in N.P.D. for 1860.	Secular Variation of Precess.in N.P.D.	Annual Proper Motion in N.P.D.	No. for reference.			
							Lacaille.	Brisbane.	Fallows or Johnson.	Greenwich or Henderson.
	° ′ ″	1800		″	″	″				
981	45 13 4·85	59·16	6	− 12·67	− 0·226	0·00	..	6913	..	1706
982	179 28 51·96	58·99	64	12·71	−13·254	+ 0·01	..	6644	J 496	H 1
983	115 46 15·71	57·91	15	12·74	− 0·397	+ 0·17	8553	..	340.J 521	1708
984	124 17 39·63	59·90	4	12·97	− 0·413	+ 0·15	8579	6922	342.J 524	1711
985	142 7 31·46	60·15	3	12·98	− 0·481	+ 0·03	8567	6921	..	..
986	117 26 23·59	56·78	5	13·12	− 0·391	+ 0·03	8601	..	..	1716
987	148 58 41·67	60·29	8	13·15	− 0·517	− 0·01	8584	6929	344.J 525	..
988	44 24 14	..	..	13·17	− 0·220	..	..	..	..	..
989	145 59 14·01	59·43	5	13·19	− 0·495	..	..	..	..	..
990	108 27 4·09	59·39	2	13·35	− 0·365	− 0·04	..	..	..	1719
991	146 14 34·53	59·52	3	13·36	− 0·491	..	..	..	..	..
992	146 19 56·47	59·38	4	13·42	− 0·489	..	..	..	..	..
993	45 20 51	..	..	13·44	− 0·221	..	..	..	..	..
994	62 28 23·19	59·31	6	13·46	− 0·271	0·00	..	..	..	1723
995	110 26 26·56	58·48	4	13·48	− 0·366	..	..	..	..	..
996	110 11 11·97	58·36	4	13·49	− 0·365	..	..	..	..	..
997	109 34 30·74	58·34	2	13·66	− 0·360	+ 0·05	..	..	..	1727
998	49 22(12·34)	..	..	13·68	− 0·232	− 0·01	..	..	..	1728
999	108 4 28·28	58·56	2	13·74	− 0·354	− 0·03	..	..	346	1731
1000	129 10 29·56	60·27	8	13·81	− 0·403	+ 0·06	8653	6962	..	..
1001	43 1(27·12)	..	..	13·87	− 0·209	+ 0·01	..	..	..	1733
1002	110 24 21·16	58·09	3	13·96	− 0·353	+ 0·05	..	..	348.J 527	1735
1003	96 22 29·37	60·54	1	13·97	− 0·326	− 0·03	..	..	..	2383*
1004	131 56 28·02	60·21	3	14·01	− 0·404	+ 0·06	8675	6970	..	..
1005	110 2 9·42	58·36	4	14·05	− 0·350	..	..	..	..	..
1006	107 47 11·56	59·51	12	14·06	− 0·345	+ 0·05	..	6976	..	1737
1007	115 33 44·28	57·36	1	14·12	− 0·359	+ 0·02	8689	..	..	1739
1008	46 37(44·92)	..	..	14·17	− 0·219	− 0·02	..	..	..	1740
1009	160 41 31·36	56·17	5	14·19	− 0·589	− 0·05	8668	6977	..	..
1010	111 45 12·65	57·59	2	14·21	− 0·349	+ 0·03	...	...	..	1741
1011	51 56 13·28	59·05	2	14·22	− 0·234	− 3·22	..	..	..	1742
1012	109 53 48·62	58·40	3	14·27	− 0·344	..	..	..	..	..
1013	120 17 9·94	60·46	2	14·29	− 0·365	..	8707	..	..	..
1014	101 56 9·49	60·43	5	14·30	− 0·328	+ 0·01	..	..	349.J 528	1745
1015	147 4 52·35	59·42	6	− 14·32	− 0·455	..	..	..	..	..

No.	No. in B.A.C.	Magnitude.	Star's Name.	Mean R.A. 1860, Jan. 1.	Mean Year and Fraction of Year	No. of Obs. of R.A.	Annual Precess. in R.A. for 1860.	Secular Variation of Precess. in R.A.	Annual Proper Motion in R.A.
				b m s	1800		s	s	s
1016	..	8	....	21 4 55·48	59·42	5	+ 4·504	— 0·0695	..
1017	7358	6	Lacaille 8718........	21 5 9·07	59·98	8	4·563	— 0·0736	..
1018	..	11·12	....	21 6 1·21	58·40	3	3·402	— 0·0141	..
1019	7368	3	64 Cygni.......... ζ	21 6 58·67	58·63	3	2·550	+ 0·0038	— 0·003
1020	7371	5·6	28 Capricorni..... φ	21 7 40	..	..	3·427	— 0·0152	+ 0·001
1021	7374	6	29 Capricorni........	21 7 59	..	..	3·329	— 0·0119	— 0·001
1022	7386	5	4 Piscis Australis...	21 9 26·35	60·20	6	3·655	— 0·0244	+ 0·001
1023	7384	6	Octantis.............	21 10 37·04	56·17	5	10·722	— 1·1295	..
1024	..	10	....	21 11 11·24	59·45	5	4·479	— 0·0715	..
1025	..	10	....	21 12 44·50	59·47	4	4·466	— 0·0716	..
1026	..	7	Lacaille 8787........	21 13 29·91	60·48	3	3·580	— 0·0218	..
1027	..	7	....	21 13 52·17	58·35	3	3·383	— 0·0140	..
1028	7407	4·5	32 Capricorni...... ι	21 14 26·76	59·82	14	3·350	— 0·0130	0·000
1029	7409	3	Pavonis............ γ	21 14 49·07	59·98	3	5·054	— 0·1208	+ 0·019
1030	7423	5	Indi.,.............. γ	21 16 14·72	59·96	3	4·336	— 0·0643	— 0·005
1031	..	8·9	....	21 18 23·84	59·41	5	4·446	— 0·0734	..
1032	..	..	....	21 18 26·24	60·38	1	3·560	— 0·0217	..
1033	7445	4	34 Capricorni..... ζ	21 18 40·03	57·14	2	3·440	— 0·0167	— 0·002
1034	..	10	....	21 20 53·41	59·45	4	4·429	— 0·0737	..
1035	..	10·11	....	21 23 4·90	59·57	4	4·426	— 0·0748	..
1036	7471	5·6	Lacaille 8833.......	21 23 13·24	60·39	1	3·827	— 0·0357	..
1037	7478	3	22 Aquarii........ β	21 24 11·15	59·63	38	3·163	— 0·0072	— 0·001
1038	..	9	....	21 25 6·74	59·54	5	4·411	— 0·0748	..
1039	..	10·11	....	21 28 39·86	59·59	4	4·402	— 0·0764	..
1040	7498	5·6	Octantis............ λ	21 28 59·60	56·34	4	10·101	— 1·1378	..
1041	7506	5·4	39 Capricorni ..... ε	21 29 14·17	58·42	2	3·371	— 0·0149	— 0·002
1042	7514	5·4	23 Aquarii........ ξ	21 30 17·71	60·76	3	3·193	— 0·0082	+ 0·004
1043	..	7	....	21 30 56·52	59·53	3	4·387	— 0·0765	..
1044	7525	4·3	40 Capricorni..... γ	21 32 19·78	58·46	15	3·322	— 0·0131	+ 0·013
1045	..	8	....	21 33 22·87	59·49	4	4·377	— 0·0773	..
1046	7538	6	Lacaille 8886........	21 34 5·25	59·90	8	3·843	— 0·0396	..
1047	..	11	....	21 35 12·79	59·44	4	4·371	— 0·0780	..
1048	7557	5	9 Piscis Australis. ι	21 36 35·84	60·36	5	3·594	— 0·0261	0·000
1049	7561	2·3	8 Pegasi............ ε	21 37 18·60	59·38	10	2·945	— 0·0006	+ 0·003
1050	7580	3	49 Capricorni..... δ	21 39 18·54	57·84	12	+ 3·304	— 0·0128	+ 0·014

No.	Mean N.P.D. 1860, Jan. 1.	Mean Year and Fraction of Year.	No. of Obs. of N.P.D.	Annual Precess. in N.P.D. for 1860.	Secular Variation of Precess.in N.P.D.	Annual Proper Motion in N.P.D.	No. for reference.			
							Lacaille.	Brisbane.	Fallows or Johnson.	Greenwich or Henderson.
	° ′ ″	1800		′	″	″				
1016	147 7 59·06	59·42	5	− 14·48	− 0·449	..	..	..	..	..
1017	148 12 23·48	59·95	9	14·50	− 0·454	+ 0·02	8718	6989	..	..
1018	109 41 32·97	58·39	4	14·55	− 0·336	..	..	..	..	..
1019	60 20 45·08	58·88	5	14·61	− 0·249	+ 0·07	..	..	..	1749
1020	111 13 48·32	57·36	1	14·65	− 0·335	− 0·07	..	..	..	1750
1021	105 45 3·13	60·53	2	14·67	− 0·325	− 0·01	..	..	350·J 529	1752
1022	122 45 17·25	60·20	6	14·75	− 0·355	+ 0·04	8761	7002	J 530	1757
1023	173 17 9·87	56·17	5	14·82	− 1·049	+ 0·19	8672	6996	..	..
1024	147 26 19·97	59·45	5	14·85	− 0·433	..	..	..	..	..
1025	147 23 40·75	59·47	4	14·95	− 0·428	..	..	..	..	..
1026	119 45 26·19	60·48	3	15·00	− 0·340	..	8787	..	..	..
1027	109 19 36·35	58·35	3	15·02	− 0·321	..	..	..	..	..
1028	107 25 41·90	59·29	16	15·05	− 0·317	− 0·02	..	..	J 532	1763
1029	155 59 44·90	59·98	3	15·07	− 0·480	− 0·83	8778	7014	353·J531	H 20
1030	145 15 43·09	60·02	5	15·15	− 0·408	− 0·04	8792	7017	J 533	..
1031	147 45 6·15	59·41	5	15·27	− 0·413	..	..	..	..	..
1032	119 24 40·87	60·38	1	15·28	− 0·329	..	..	..	..	..
1033	113 0 55·30	57·14	2	15·29	− 0·318	− 0·02	8815	..	356·J 534	1775
1034	147 46 13·54	59·45	4	15·41	− 0·406	..	..	..	..	..
1035	148 0 2·59	59·57	5	15·53	− 0·401	..	..	..	..	..
1036	131 47 38·33	60·39	1	15·54	− 0·346	+ 0·20	8833	7036	..	..
1037	96 11 5·32	59·12	95	15·60	− 0·283	0·00	..	7040	357·J 535	1777
1038	147 59 49·37	59·55	6	15·64	− 0 396	..	..	..	..	..
1039	148 20 15·31	59·58	5	15·84	− 0·387	..	..	..	..	..
1040	173 21 23·81	56·90	5	15·85	− 0·895	+ 0·10	8798	7042	..	..
1041	110 5 27·72	58·42	2	15·87	− 0·294	0·00	..	..	J 536	1787
1042	98 28 48·09	60·76	3	15·93	− 0·276	+ 0·04	..	7055	358·J 537	1788
1043	148 22 7·42	59·54	4	15·96	− 0·381	..	..	..	..	..
1044	107 17 33·05	58·43	23	16·04	− 0·284	+ 0·03	..	..	359·J 538	1790
1045	148 31 59·11	59·51	4	16·09	− 0·374	..	..	..	..	..
1046	134 7 46·24	59·90	8	16·13	− 0·326	+ 0·02	8886	7068	..	..
1047	148 41 18·82	59·45	4	16·18	− 0·370	..	..	..	..	..
1048	123 39 44·35	60·36	5	16·26	− 0·300	+ 0·11	8901	7074	J 541	1801
1049	80 45 54·41	59·08	21	16·29	− 0·244	0·00	..	..	..	1804
1050	106 45 37·53	58·31	21	− 16·40	− 0·271	+ 0·28	..	..	364·J 542	1811

No.	No. in B.A.C.	Magnitude.	Star's Name.	Mean R.A. 1860, Jan. 1.	Mean Year and Fraction of Year.	No. of Obs. of R.A.	Annual Precess. in R.A. for 1860.	Secular Variation of Precess. in R.A.	Annual Proper Motion in R.A.
				h m s	1800		s	s	s
1051	7583	5·4	10 Piscis Australis θ	21 39 30·50	60·45	4	+ 3·544	− 0·0240	+ 0·002
1052	..	8	....	21 40 22·46	59·47	4	4·343	− 0·0790	..
1053	7613	3	Gruis............ γ	21 45 26·30	60·07	10	3·651	− 0·0311	+ 0·002
1054	7618	5	51 Capricorni..... μ	21 45 39·56	57·93	4	3·259	− 0·0112	+ 0·021
1055	7627	5·6	16 Pegasi...........	21 46 41·65	58·43	4	2·725	+ 0·0052	+ 0·001
1056	7633	5	Indi............... δ	21 48 21·64	60·48	1	4·135	− 0·0666	− 0·005
1057	7634	5	Indi............... κ¹	21 48 34·30	59·74	1	4·312	− 0·0820	+ 0·009
1058	7657	5·6	12 Piscis Australis η	21 52 47·07	60·18	10	3·465	− 0·0219	+ 0·003
1059	..	7	Lacaille 9012	21 56 37·80	60·46	3	3·454	− 0·0219	..
1060	7688	·3	34 Aquarii........ α	21 58 35·49	58·87	12	3·084	− 0·0042	− 0·003
1061	7691	4	33 Aquarii........ ι	21 58 52·37	58·38	6	3·247	− 0·0113	− 0·001
1062	7692	2	Gruis........ ...... α	21 59 23·49	59·83	17	3·808	− 0·0459	+ 0·011
1063	..	9	....	22 0 29·46	58·42	3	3·281	− 0·0133	..
1064	7713	6	Octantis.......... C	22 3 34·83	58·69	50	14·262	− 3·7285	− 0·030
1065	7725	5·6	Octantis.......... ε	22 4 4·84	56·32	5	7·224	− 0·6251	+ 0·027
1066	..	8	....	22 5 29·96	58·42	3	3·272	− 0·0129	..
1067	7748	6	Lacaille 9061	22 6 4·71	59·65	1	3·647	− 0·0363	+ 0·051
1068	7756	5	Gruis............ μ¹	22 7 10·03	60·25	4	3·642	− 0·0362	− 0·001
1069	7763	5·6	Gruis............ μ²	22 8 0·19	57·70	1	3·644	− 0·0365	− 0·005
1070	..	11	....	22 8 46·20	59·52	·5	4·176	− 0·0843	..
1071	7767	·3	Tucanæ.......... α	22 8 52·65	59·76	2	4·194	−. 0·0862	− 0·007
1072	7771	6	42 Aquarii...........	22 9 17	..	..	3·221	− 0·0105	..
1073	7773	4·5	43 Aquarii........ θ	22 9 26·61	59·71	15	3·165	− 0·0077	+ 0·006
1074	..	10	....	22 9 41·01	59·51	4	4·181	− 0·0856	..
1075	7781	6	45 Aquarii...........	22 11 29·71	57·52	1	3·224	− 0·0107	+ 0·009
1076	..	10	....	22 12 28·75	58·43	4	3·252	− 0·0123	..
1077	..	12	....	22 16 41·02	59·65	5	4·128	− 0·0858	..
1078	7806	6	50 Aquarii...........	22 16 56·93	58·12	4	3·220	− 0·0107	+ 0·004
1079	7808	5	Tucanæ............ δ	22 17 19·70	59·38	6	4·354	− 0·1129	+ 0·006
1080	..	12	....	22 18 44·62	59·56	6	4·116	− 0·0864	..
1081	7828	4	Gruis............ δ¹	22 20 53·10	60·25	6	3·617	− 0·0391	− 0·007
1082	..	11	....	22 21 16·43	59·51	4	4·089	− 0·0856	..
1083	7830	5	Gruis............ δ²	22 21 22·77	59·77	1	3·619	− 0·0394	− 0·006
1084	7832	3·4	55 Aquarii......... ζ	22 21 37·19	60·51	1	3·079	− 0·0034	+ 0·009
1085	7840	5·4	57 Aquarii......... σ	22 23 14·06	59·03	6	+ 3·182	− 0·0089	− 0·004

No.	Mean N.P.D. 1860, Jan. 1.	Mean Year and Fraction of Year.	No. of Obs. of N.P.D.	Annual Precess. in N.P.D. for 1860.	Secular Variation of Precess.in N.P.D.	Annual Proper Motion in N.P.D.	No. for reference.			
							Lacaille.	Brisbane.	Fallows or Johnson.	Greenwich or Henderson.
	° ′ ″	1800		″	″	″				
1051	121 32 38·02	60·45	4	− 16·40	− 0·290	0·00	8917	7082	J 543	1812
1052	148 57 0·90	59·47	4	16·45	− 0·356	··	··	··	··	··
1053	128 1 16·00	60·01	9	16·70	− 0·288	+ 0·02	8951	7094	365·J 544	··
1054	104 12 31·40	58·04	5	16·71	− 0·256	− 0·02	··	··	J 545	1822
1055	64 43 56·18	58·55	6	16·76	− 0·211	+ 0·01	··	··	··	1824
1056	145 39 19·91	60·48	1	16·84	− 0·321	+ 0·01	8962	7100	J 546	··
1057	149 40 38·58	59·74	1	16·85	− 0·334	− 0·09	8959	7101	··	··
1058	119 7 25·03	60·14	9	17·04	− 0·259	− 0·03	··	7112	··	1832
1059	119 6 42·23	60·54	3	17·22	− 0·251	··	9012	··	··	··
1060	90 59 54·23	59·18	33	17·31	− 0·220	+ 0·02	··	7129	367·J 550	1840
1061	104 32 49·62	58·51	11	17·32	− 0·231	+ 0·07	··	··	368·J 551	1842
1062	137 38 11·89	59·95	18	17·34	− 0·272	+ 0·15	9021	7130	369·J 552	H 53
1063	107 22 19·76	58·42	3	17·38	− 0·231	··	··	··	··	··
1064	176 40 24·79	58·46	57	17·52	− 1·004	− 0·08	8924	7119	J 549	H 7
1065	171 7 58·44	56·32	5	17·54	− 0·502	+ 0·07	9010	7134	··	··
1066	107 18 42·71	58·42	3	17·60	− 0·221	··	··	··	··	··
1067	132 2 35·08	59·65	1	17·63	− 0·247	+ 0·75	9061	7144	J 553	··
1068	132 2 28·84	60·25	4	17·67	− 0·243	− 0·03	9069	7146	J 554	··
1069	132 19 20·69	57·70	1	17·71	− 0·242	+ 0·11	9075	7148	··	··
1070	150 32 3·57	59·52	5	17·74	− 0·276	··	··	··	··	··
1071	150 57 18·58	59·76	2	17·75	− 0·277	+ 0·04	9074	7149	374·J 555	H 28
1072	103 31 40·34	57·67	1	17·76	− 0·211	··	··	··	375	··
1073	98 28 43·66	59·67	35	17·77	− 0·206	+ 0·03	··	7151	376·J 556	1860
1074	150 48 57·05	59·51	4	17·78	− 0·274	··	··	··	··	··
1075	104 0 15·42	57·52	1	17·85	− 0·207	− 0·05	··	··	··	1863
1076	106 35 23·78	58·43	4	17·88	− 0·207	··	··	··	··	··
1077	151 5 32·67	59·65	5	18·05	− 0·254	··	··	··	··	··
1078	104 14 15·27	58·12	4	18·06	− 0·196	− 0·04	··	··	··	1873
1079	155 40 36·11	59·45	8	18·08	− 0·267	− 0·03	9114	7163	J 559	··
1080	151 17 13·93	59·58	5	18 13	− 0·249	··	··	··	··	··
1081	134 12 33·22	60·34	5	18·21	− 0·213	+ 0·01	9138	7172	J 560	··
1082	151 13 22·93	59·51	3	18·22	− 0·241	··	··	··	··	··
1083	134 27 50·74	59·77	1	18·23	− 0·212	+ 0·07	9140	7173	J 561	··
1084	90 44 6·93	60·51	1	18·24	− 0·179	− 0·03	··	··	381·J 562	1878
1085	101 23 34·46	59·07	12	− 18·30	− 0·182	− 0·05	··	··	J 563	1882

No.	No. in B.A.C.	Magnitude.	Star's Name.	Mean R.A. 1860, Jan. 1.	Mean Year and Fraction of Year.	No. of Obs. of R.A.	Annual Precess. in R.A. for 1860.	Secular Variation of Precess. in R.A.	Annual Proper Motion in R.A.
				h m s	1800		s	s	s
1086	7842	4	17 Piscis Australis β	22 23 32·14	60·43	3	+ 3·428	− 0·0250	+ 0·002
1087	..	10	....	22 23 38·25	58·44	4	3·228	− 0·0116	..
1088	..	8	....	22 23 58·17	59·48	4	4·076	− 0·0866	..
1089	7849	6	58 Aquarii............	22 24 15·75	57·82	1	3·184	− 0·0090	+ 0·006
1090	..	11	....	22 25 45·08	59·65	7	4·064	− 0·0869	..
1091	..	9·10	....	22 27 29·70	59·57	5	4·049	− 0·0867	..
1092	7868	4·3	62 Aquarii......... η	22 28 9·70	60·05	11	3·080	− 0·0032	+ 0·003
1093	7884	5	63 Aquarii......... κ	22 30 30·28	59·99	3	3·116	− 0·0052	− 0·007
1094	..	9	....	22 30 58·32	59·46	4	4·024	− 0·0872	..
1095	..	9	....	22 31 16·29	58·44	3	3·213	− 0·0111	..
1096	7886	5	Octantis............ β	22 31 29·05	56·27	6	6·708	− 0·6767	− 0·034
1097	7887	6·7	Gruis................	22 31 35·91	57·70	1	3·679	− 0·0494	..
1098	7898	4	18 Piscis Australis ε	22 32 54·23	60·00	3	3·333	− 0·0198	− 0·007
1099	7904	3	Gruis .... β	22 34 17·40	60·21	7	3·607	− 0·0439	+ 0·012
1100	7908	3·4	42 Pegasi........... ζ	22 34 28·84	58·98	3	2·985	+ 0·0022	+ 0·001
1101	7909	6	19 Piscis Australis..	22 34 34·33	60·16	7	3·354	− 0·0217	+ 0·003
1102	7921	6	67 Aquarii............	22 35 55·49	58·72	1	3·137	− 0·0064	0·000
1103	..	7	....	22 36 49·98	58·42	4	3·202	− 0·0108	..
1104	7925	3	Gruis............... η	22 37 0·80	59·89	4	3·730	− 0·0581	− 0·004
1105	7946	4	Gruis............... ε	22 40 4·52	59·84	12	3·658	− 0·0523	+ 0·003
1106	7952	6	70 Aquarii............	22 41 8·02	58·30	3	3·162	− 0·0082	+ 0·006
1107	7954	4	71 Aquarii....... τ²	22 42 10·59	56·85	1	3·186	− 0·0100	− 0·004
1108	..	7	....	22 42 58·70	58·42	4	3·189	− 0·0101	..
1109	7966	5·4	22 Piscis Australis γ	22 44 44·03	59·94	10	3·359	− 0·0245	− 0·004
1110	7970	4	73 Aquarii......... λ	22 45 18·50	56·72	3	3·135	− 0·0064	− 0·006
1111	7980	3	76 Aquarii......... δ	22 47 12·98	56·78	2	3·196	− 0·0111	− 0·007
1112	..	7·8	....	22 47 59·37	58·41	3	3·179	− 0·0099	..
1113	7992	1·2	24 Piscis Australis α	22 49 54·36	59·56	40	3·308	− 0·0212	+ 0·022
1114	..	1·1	....	22 51 19·21	58·45	4	3·172	− 0·0095	..
1115	8008	5	Gruis............ ζ	22 52 35·51	59·87	6	3·596	− 0·0537	− 0·011
1116	8016	6	81 Aquarii............	22 54 7·03	57·52	1	3·124	− 0·0057	+ 0·002
1117	..	9·10	....	22 54 18·04	58·45	3	3·165	− 0·0091	..
1118	8020	6	82 Aquarii............	22 55 16·35	58·65	1	3·120	− 0·0054	+ 0·001
1119	8031	5·4	4 Piscium......... β	22 56 45·16	60·74	2	3·053	0·0000	+ 0·001
1120	8034	2	54 Pegasi.......... α	22 57 47·30	58·77	4	+ 2·980	+ 0·0056	+ 0·003

No.	Mean N.P.D. 1860, Jan. 1.	Mean Year and Fraction of Year.	No. of Obs. of N.P.D.	Annual Precess. in N.P.D. for 1860.	Secular Variation of Precess. in N.P.D.	Annual Proper Motion in N.P.D.	No. for reference.			
							Lacaille.	Brisbane.	Fallows or Johnson.	Greenwich or Henderson.
1086	123 3 44·91	1800 60·42	2	− 18·31	− 0·196	+ 0·07	9162	7176	J 564	1883
1087	105 56 57·26	58·44	4	18·30	− 0·185	··	··	··	··	··
1088	151 32 8·91	59·41	4	18·32	− 0·234	··	··	··	··	··
1089	101 37 16·52	57·82	1	18·33	− 0·181	− 0·01	··	··	··	1887
1090	151 40 14·45	59·65	7	18·38	− 0·229	··	··	··	··	··
1091	151 43 35·07	59·57	5	18·44	− 0·224	··	··	··	··	··
1092	90 50 16·57	59·82	22	18·47	− 0·168	+ 0·06	··	··	382.J 566	1892
1093	94 56 56·37	59·97	3	18·55	− 0·165	+ 0·11	··	··	383	1894
1094	151 57 40·79	59·46	4	18·56	− 0·214	··	··	··	··	··
1095	105 32 22·26	58·44	4	18·57	− 0·169	··	··	··	··	··
1096	172 6 46·82	56·27	6	18·58	− 0·361	0·00	9165	7186	J 567	H 12
1097	140 19 24·75	57·70	1	18·58	− 0·194	+ 0·23	9200	7188	··	··
1098	117 46 21·41	60·00	3	18·62	− 0·172	+ 0·06	9206	7193	J 568	··
1099	137 36 54·71	60·21	7	18·67	− 0·184	+ 0·04	9211	7194	384.J 569	H 54
1100	79 53 54·03	59·31	9	18·68	− 0·151	0·00	··	··	··	1900
1101	120 5 28·85	60·16	7	18·68	− 0·170	+ 0·09	··	7197	··	1901
1102	97 41(39·71)	··	··	18·72	− 0·156	− 0·05	··	··	385	1903
1103	105 20 41·45	58·42	4	18·75	− 0·158	··	··	··	··	··
1104	144 14 6·11	59·86	5	18·75	− 0·185	0·00	9223	7203	J 570	··
1105	142 3 6·32	59·84	13	18·85	− 0·174	+ 0·11	9249	7212	386.J 571	··
1106	101 17 36·77	58·30	3	18·88	− 0·148	− 0·04	··	··	··	1907
1107	104 19 48·63	56·85	1	18·91	− 0·147	+ 0·02	··	··	··	1908
1108	104 47 55·29	58·42	4	18·93	− 0·146	··	··	··	··	··
1109	123 37 0·89	59·94	10	18·98	− 0·150	+ 0·04	9287	7218	J 572	1912
1110	98 19 24·22	58·11	11	19·00	− 0·138	− 0·03	··	··	388.J 573	1913
1111	106 33 51·09	56·78	2	19·05	− 0·137	0·00	··	··	389.J 574	1917
1112	104 29 8·08	58·41	3	19·07	− 0·136	··	··	··	··	··
1113	120 21 47·28	58·99	123	19·13	− 0·137	+ 0·18	9314	7225	391.J 575	1920
1114	104 12 53·58	58·45	3	19·16	− 0·129	··	··	··	··	··
1115	143 30 13·91	59·87	6	19·19	− 0·144	− 0·07	9322	7229	J 576	··
1116	97 48 42·47	57·52	1	19·23	− 0·121	− 0·06	··	··	393	1923
1117	103 49 6·36	58·45	3	19·24	− 0·123	··	··	··	··	··
1118	97 19 29·38	58·65	1	19·26	− 0·119	− 0·01	··	··	··	1925
1119	86 55 58·62	60·74	2	19·30	− 0·113	+ 0·02	··	··	··	1927
1120	75 32 49·52	59·44	7	− 19·32	− 0·108	+ 0·02	··	7239	··	1929

No.	No. in B.A.C.	Magnitude.	Star's Name.	Mean R.A. 1860, Jan. 1.	Mean Year and Fraction of Year.	No. of Obs. of R.A.	Annual Precess. in R.A. for 1860.	Secular Variation of Precess. in R.A.	Annual Proper Motion in R.A.
				h m s	1800		s	s	s
1121	8035	6·5	83 Aquarii........ h¹	22 57 51·70	58·83	2	+ 3·125	− 0·0059	+ 0·013
1122	8043	5	Gruis............... θ	22 58 58·68	60·01	9	3·415	− 0·0358	− 0·003
1123	8060	6	5 Piscium........ A	23 1 30·60	60·81	2	3·064	− 0·0006	+ 0·001
1124	8067	5	Gruis.............. ι	23 2 25·06	59·91	6	3·416	− 0·0380	+ 0·007
1125	8072	6	Octantis........... τ	23 5 3·30	58·52	104	13·456	− 7·3697	+ 0·037
1126	8085	4·5	90 Aquarii........ φ	23 7 4·27	59·17	2	3·109	− 0·0046	+ 0·001
1127	8093	5	Lacaille 9412.......	23 8 30·59	59·91	5	3·651	− 0·0791	+ 0·017
1128	8098	4	Tucanæ............ γ	23 9 13·82	60·04	3	3·561	− 0·0650	− 0·012
1129	8105	4	6 Piscium......... γ	23 9 54·47	59·81	11	3·059	+ 0·0004	+ 0·047
1130	8109	5·4	93 Aquarii....... ψ²	23 10 37·50	57·60	4	3·122	− 0·0061	0·000
1131	8116	5	95 Aquarii..... ψ³	23 11 40·59	56·97	3	3·123	− 0·0063	− 0·002
1132	8119	6·5	96 Aquarii.........	23 12 8·36	59·22	3	3·101	− 0·0038	+ 0·011
1133	8157	5·6	Lacaille 9463........	23 17 18·75	60·03	5	3·462	− 0·0586	+ 0·005
1134	8169	5·4	8 Piscium......... κ	23 19 45·34	59·63	7	3·070	− 0·0001	+ 0·005
1135	8186	6	Lacaille 9495.......	23 22 59·86	60·49	1	3·271	− 0·0310	..
1136	8201	5	Sculptoris......... β	23 25 27·21	60·56	3	3·232	− 0·0262	+ 0·004
1137	8210	5	Phœnicis............ ι	23 27 31·98	60·09	5	3·250	− 0·0312	− 0·001
1138	8218	6	16 Piscium..........	23 29 14·65	60·12	3	3·068	+ 0·0009	− 0·006
1139	8230	5·6	Phœnicis............ θ	23 31 56·03	59·92	5	3·250	− 0·0357	− 0·017
1140	8233	4·5	17 Piscium........ ι	23 32 45·09	57·97	4	3·059	+ 0·0029	+ 0·025
1141	8243	5	18 Piscium........ λ	23 34 54·18	58·66	8	3·069	+ 0·0010	− 0·011
1142	8254	6	Lacaille 9574........	23 36 30·29	59·88	5	3·213	− 0·0329	..
1143	8262	6	19 Piscium..........	23 39 14·30	57·75	2	3·066	+ 0·0021	+ 0·002
1144	8271	6	20 Piscium..........	23 40 44·66	56·70	2	3·079	− 0·0010	+ 0·002
1145	8275	4·5	Sculptoris.......... δ	23 41 37·62	59·76	11	3·131	− 0·0162	+ 0·009
1146	8281	6	21 Piscium..........	23 42 17·32	57·83	1	3·072	+ 0·0011	+ 0·003
1147	8290	5	Octantis.......... γ¹	23 43 45·29	56·40	6	3·821	− 0·3500	− 0·038
1148	8295	6	22 Piscium..........	23 44 47·76	58·58	1	3·069	+ 0·0022	+ 0·004
1149	..	8·9	Lalande 46854......	23 47 28·22	59·68	1	3·075	− 0·0002	..
1150	8312	6	26 Piscium..........	23 47 58·06	60·05	1	3·064	+ 0·0045	+ 0·005
1151	8319	5	Octantis.......... γ³	23 49 44·96	56·44	6	3·555	− 0·3160	− 0·018
1152	8323	5	Tucanæ............ η	23 50 12·41	59·85	3	3·195	− 0·0680	+ 0·015
1153	8328	5·6	27 Piscium.	23 51 30·30	57·64	2	3·076	− 0·0008	− 0·008
1154	8331	4	28 Piscium........ ω	23 52 7·40	59·48	12	3·067	+ 0·0046	+ 0·010
1155	8334	5	Tucanæ............ ε	23 52 36·66	59·93	6	+ 3·171	− 0·0711	− 0·001

1139. The companion was observed in 1860. The Right Ascension was 0·06 greater than that of the principal Star.

No.	Mean N.P.D. 1860, Jan. 1.	Mean Year and Fraction of Year.	No. of Obs. of N.P.D.	Annual Precess. in N.P.D. for 1860.	Secular Variation of Precess.in N.P.D.	Annual Proper Motion in N.P.D.	No. for reference.			
							Lacaille.	Brisbane.	Fellows or Johnson.	Greenwich or Henderson.
	° ′ ″	1800		′	′	′				
1121	98 26 55·18	58·83	2	− 19·32	− 0·114	− 0·06	..	..	395	1930
1122	134 16 31·47	60·01	9	19·35	− 0·123	+ 0·11	9366	7244	J 577	..
1123	88 38 1·57	60·81	2	19·41	− 0·105	− 0·15	..	..	..	1935
1124	136 0 15·02	60·00	7	19·43	− 0·116	+ 0·02	9382	7252	397·J 580	..
1125	178 14 55·73	58·25	126	19·48	− 0·458	− 0·02	9225	7241	J 578	H 4
1126	96 48 10·93	59·30	10	19·52	− 0·095	+ 0·19	..	..	399.J 582	1942
1127	152 45 47·77	59·92	4	19·55	− 0·110	+ 0·03	9412	7266	..	..
1128	149 0 9·14	60·04	3	19·57	− 0·106	− 0·04	9420	7267	400.J 583	..
1129	87 28 55·42	59·61	26	19·58	− 0·088	+ 0·01	..	7269	..	1947
1130	99 56 45·33	57·60	4	19·59	− 0·089	+ 0·02	..	..	401.J 584	1948
1131	100 22 31·41	56·97	3	19·61	− 0·087	− 0·01	..	..	J 586	1952
1132	95 53 18·98	59·22	3	19·62	− 0·085	+ 0·01	..	..	402	1953
1133	147 37 0·60	59·90	4	19·71	− 0·085	− 0·16	9463	7285	..	..
1134	89 30 37·55	59·51	36	19·75	− 0·070	+ 0·12	..	..	407	1962
1135	132 45 26·56	60·17	2	19·79	− 0·068	+ 0·14	9495	7296	..	..
1136	128 35 30·48	60·13	4	19·83	− 0·062	− 0·02	9513	7300	411.J 589	..
1137	133 23 18·96	60·05	6	19·85	− 0·058	− 0·03	9523	7304	J 591	..
1138	88 40 27·83	60·12	3	19·88	− 0·051	− 0·06	..	..	..	1973
1139	137 24 52·28	59·98	8	19·91	− 0·049	0·00	9543	7315	J 592	..
1140	85 7 55·69	59·07	23	19·91	− 0·044	+ 0·45	..	..	..	1978
1141	88 59 24·50	58·45	10	19·94	− 0·040	+ 0·17	..	..	..	1985
1142	135 51 36·27	59·72	..	19·95	− 0·039	..	9574	..	..	..
1143	87 17 23·04	57·75	2	19·97	− 0·032	− 0·02	..	..	..	1990
1144	93 32 22·09	56·70	2	19·99	− 0·029	+ 0·01	..	..	416	1992
1145	118 54 15·16	59·33	16	19·99	− 0·028	+ 0·10	9603	7330	417.J 597	1994
1146	89 42 3·42	58·79	7	20·00	− 0·026	+ 0·08	..	..	..	1995
1147	172 47 48·32	56·40	6	20·01	− 0·031	+ 0·02	9607	7334	J 598	H 10
1148	87 50 50·03	58·58	1	20·01	− 0·021	− 0·02	..	..	..	1996
1149	92 44 28·40	59·68	1	20·02	− 0·016	..	..	..	..	..
1150	83 42 26·60	60·05	5	20·03	− 0·015	− 0·02	..	..	..	1999
1151	172 56 53·52	56·43	6	20·04	− 0·015	+ 0·01	9651	7350	418.J 599	H 9
1152	155 4 31·33	59·85	2	20·04	− 0·011	+ 0·02	9661	7352	J 600	..
1153	94 19 56·70	57·64	2	20·04	− 0·008	+ 0·12	..	..	419.J 601	2003
1154	83 54 42·51	59·58	27	20·04	− 0·007	+ 0·13	..	..	..	2004
1155	156 21 21·09	59·95	5	− 20·05	− 0·006	− 0·03	9678	7360	420.J 602	..

No.	No. in B.A.C.	Magnitude.	Star's Name.	Mean R.A. 1860, Jan. 1.	Mean Year and Fraction of Year.	No. of Obs. of R.A.	Annual Precess. in R.A. for 1860.	Secular Variation of Precess. in R.A.	Annual Proper Motion in R.A.
				h m s	1800		s	s	s
1156	8346	5·6	29 Piscium...........	23 54 38·76	58·02	2	+ 3·074	— 0·0004	— 0·002
1157	8349	5	30 Piscium..........	23 54 46·74	56·73	3	3·076	— 0·0020	+ 0·002
1158	8368	5	33 Piscium..........	23 58 10·13	56·70	2	3·073	— 0·0016	— 0·002
1159	..	..	....	23 59 10·65	60·48	1	+ 3·072	+ 0·0011	..

No.	Mean N.P.D. 1860, Jan. 1.	Mean Year and Fraction of Year.	No. of Obs. of N.P.D.	Annual Precess. in N.P.D. for 1860.	Secular Variation of Precess. in N.P.D.	Annual Proper Motion in N.P.D.	No. for reference.			
							Lacaille.	Brisbane.	Fallows or Johnson.	Greenwich or Henderson.
	° ′ ″	1800		″	″	″				
1156	93 48 23·40	58·02	2	− 20·05	− 0·002	+ 0·01	..	..	421.J 603	2008
1157	96 47 31·31	56·73	3	20·05	− 0·002	+ 0·04	..	..	J 604	2009
1158	96 29 26·41	56·73	3	20·06	+ 0·005	− 0·03	..	..	424.J 606	2019
1159	91 27 7·63	60·48	1	− 20·06	+ 0·007	..	..	..	..	..

www.ingramcontent.com/pod-product-compliance
Lightning Source LLC
Chambersburg PA
CBHW032358020726
47499CB00008B/2800